"Our actions now will define the future of our society (…) it is no longer enough to simply act. We must act quickly, systematically and together."
Martin R. Stuchtey, Sandrine Dixson-Declève, Janez Potočnik, et al.
(A System Change Compass, October 2020, p.ii)

"Change is unavoidable (…) it has to do a lot with the hard laws of physics. We need to change (…) The whole economy has to change (…) You can simply not manage the 21st century with short-termism anymore (…) If there is anything from which we can learn, it's absolutely nature; because nature, it's the biggest and the best circular economy (…) We belong to (…) nature. And that's why we have to behave like we are part of (…) nature (…) Environment and economy are two sides of the same coin."
Janez Potočnik
(New environmentalism and the circular economy: Janez Potočnik at TEDxFlanders, 04.04.2014)

"... the process of (…) literally transforming mind into matter. **Emotions** are the nexus between matter and mind, going back and forth between the two and influencing both."
Candance Pert
(Molecules of Emotion, 1997, p.189; note: bold Authors)

"The most beautiful thing we can experience is the mysterious. It is the source of all true art and science. He to whom **this emotion** is a stranger, who can no longer pause to wonder and stand rapt in awe, is as good as dead: his eyes are closed."
Albert Einstein
(What I believe; Forum and Century 84; October 1930, No.4, 194-194; note: bold Authors)

The Posionites & the Global Green Deal

**An Action Comedy Novel
inspired by the report *A System Change Compass*,
co-authored by The Club of Rome and SYSTEMIQ**

**Written by
George Hohbach & Ehrengard Hohbach**

**Illustrations & photo by
George Hohbach**

**Music & lyrics by
George Hohbach**

**Arrangement and sheet music by
Alfred Huff**

In anticipation of the book, George and Ehrengard planted 100 trees in Iceland in February of 2021 with the Environmental Charity *One Tree Planted.*

Bibliographical Information of the Deutsche Nationalbibliothek
This publication is listed in the Deutsche Nationalbibliographie of the
Deutsche Nationalbibliothek; detailed bibliographical information
can be accessed under http: //dnb.d-nb.de

Printing and Production: BoD – Books on Demand, Norderstedt

Illustrations, photo and cover design by George Hohbach

ISBN: 978-3-7534-6588-3

MIX
Papier aus verantwortungsvollen Quellen
Paper from responsible sources
FSC® C105338
FSC
www.fsc.org

Contents

PART 1

THE POSIONITES
& THE GLOBAL GREEN DEAL

CHAPTER 1

The universe, seriously, is vast. Like infinitely vast. You gotta give it to the universe, this is quite an achievement. Like, the universe for instance, has infinitely many event points in its so-called spacetime continuum. (Most event points consist of nothing visibly actually going on, so don't expect to find too many nightclubs out there.) Thing is, the universe, which the Greeks called cosmos, and means "order" (yeah!), is full of galaxies, even more stars and lots of other cool, creativity-friendly stuff like nebulae, planets, rocks, stones, dust-like particles. Well, I think you get the picture. So, somewhere in all that, is our Milky Way Galaxy, and somewhere in that somewhere, sort of in a small spiral arm of our galaxy, is our solar system. The star of our solar system literally is a star, the sun. The sun, shining bright, happy and enormously confident, and around it, we have the planets, orbiting it as the central, radiating energy source. The blue round planet that we live on, is the third from the sun, right after Venus and Mercury (in case you wanna include that on your stationary to make your address stand out). Countless satellites are circling our home planet, like the satellites of news channels like, uh, that one: Super Global Broadcasting News Network, or in short SGBNN.

One of SGBNN's mega-charged "Good Morning Planet Earth" news anchors, Harry Pearl, was just spreading exciting information around the globe. More than ever, he totally looked like he was performing at his very professional best.

"Good morning, everyone!" he said with a spectacularly uplifting smile. "We are approaching a historic moment here on planet Earth. World leaders are going to ratify a binding Global Green Deal in Sweden's capital, Stockholm, in 3 days."

Every viewer around the world could see, Harry Pearl's level of commitment to the news was at a record high. Because the news was pretty darn good today. Harry turned to his colleague, Doris Pearl, who sat right next to him at the news desk. That's the kinda random stuff that can happen in a universe with infinite possibilities. You get born as Harry Pearl and 30 years later, you sit next to another reporter whose last name was also Pearl. So, Harry and Doris with her perfectly styled hair, were known not as the dream team, but as the *Pearl Team* of morning television.

Anyhow, as the words GLOBAL GREEN DEAL whooshingly popped up above an image of planet Earth on TV screens around the globe, Doris fired off more of the good news.

"The historic Global Green Deal is largely based on an influential European report from 2020 on how to effectively implement Green Deals and is meant to do more than just treat the symptoms of the man-made environmental crisis. The ground-breaking deal will be announced by two young climate activists…"

With another swift, attention-stimulating woosh and green flash, the images of two young climate activists, a girl and a boy, both around the age of 13, appeared on TV screens worldwide.

Doris beamed with joy as she continued reporting: "The two brave and smart young thought leaders will convey to the world how this deal of a Nature-based balanced concept, will make our relationship with Nature positive again!"

As soon as Doris had uttered the word "positive", the entire news room was filled with a bright shining light. It was a strong, empowering energy that, for a long second, suddenly emanated from the Pearl Team.

Up in space, the news satellites of SGBNN simultaneously began to radiate an intense bright light, and six of them, orbiting and spread around the planet, sent a powerful beam of light down to Earth. Into the middle of the USA, to Europe, the Middle East, China, India and Africa. Let's see what happened down there.

CHAPTER 2

ISAAC AL

AGE: 14

COUNTRY: USA

MOTTO: I dare and love to be myself and sooo alive! WE have the right to feel amazing, dream wildly, laugh loudly, think with Nature & Thrive, Thrive, Thrive!

THE MIDWEST OF RURAL AMERICA

Fourteen-year-old Isaac Al, was chasing down a lonely road in the Midwest of rural America. With all his might, that is the total weight of his right foot on the gas pedal, Isaac made sure his car was "cornering like it's on rails". That's what he thought of at this very moment especially after he had seen the classic romantic comedy *Pretty Woman.* While watching the movie Pretty Woman and eating popcorn with his family, Julia Roberts, driving in a sexy Lotus with Richard Gere through Beverly Hills, had made this cool comment about his even cooler car, "It corners like it's on rails." That brought the Lotus onto Isaac and his dad's radar big time.

Isaac had a pretty cool astronaut as his dad. Once Isaac's dad had come back down to earth after a space trip, he had everybody move away from the big city to the countryside. Because after Isaac's dad had seen earth and its tiny, precious atmosphere from space, he knew he wanted his kids to experience the beauty of Nature as much as possible. But Isaac's dad was also cool for another reason, as like Isaac, he, too, was into cars. So, soon after they had seen Pretty Woman, as his mom and sister had insisted on that film, father and son had bought an old Lotus on eBay to butch things up a bit. Since both Isaac and his dad loved Star Wars movies, and since its creator, George Lucas, in his youth had also loved fast cars, it was sorta logical for both of them to buy a fast, sexy car. On top of that, Mr. Universe and kick-ass, former governor of California, Arnold Schwarzenegger, owned a Humvee that ran on hydrogen in order to be eco-intelligent. So, adding all this up led to their Lotus displaying the green slogan "Physics is awesome!" on its doors, being driven by Isaac and running very fast on hydrogen, down this lonely road. George Lucas had once said that the force is "a way of seeing" and "being with life". Right now, Isaac felt that he was one with the road, one with the beautiful landscape around him, one with the gorgeous sunrise and one with the fact that he was winning the car race with his buddies from school.

This oneness with all was interrupted by his cell phone's ring tone. It was dad!???

"Hey, dad?!" Isaac said totally surprised.

"Hi, Isaac!"

"How come you can call me, I thought they didn't let you call home from the space station?"

"Well, you know, I don't like to accept `No´ as an answer. Say `Hi´ to mom and your sister. I couldn't reach them this morning. They're apparently already too busy. So, what's up with you. On your way to school?"

"Sorta, yeah."

"Great. So, since you're moving mass through the countryside, what's F=ma to do with it?"

"F=ma. That's Newton's second law of motion. Come on, dad! I can't give you an A for asking such a simple question!" Isaac laughed while taking a corner at top speed.

"Is that tires screeching?"

"No, must be some problem with the connection," Isaac replied quickly.

"Sure. So, one last physics question. What's Einstein's true legacy?"

"Obviously his groundbreaking views on symmetry!"

"What's that?"

"That the laws of Nature are the same for all observers independent of their state of motion. Accordingly, the laws of Nature are the same everywhere at all times.”

“Awesome. So, this means you don’t have to speed through corners to appreciate the amazing elegance of Nature.”

“Dad, I’m a safe driver! You know me!”

“I know, but you’ve got your mother’s competitive spirit.”

“That’s why I love you both. Have fun, on the ISS dad! Do some great science up there so that our politicians fully wake up to reality.”

“I will. Love you, too.”

With a smile, Isaac, ended the connection. Instantly, he was one again with the road, the speed and the fact that he was way ahead of his buddies and winning this race for certain. At least that’s what Isaac thought, until he heard a loud bang come from his engine. His instant “Darn it!” was followed a split second later by a subsequent “Shit!”, as the car began to stutter.

His buddies in the car behind him weren’t shocked at all, when Isaac unexplainably slowed down in front of them. They were just so frigging excited that a window of opportunity had magically opened up. Five seconds later, they sped by Isaac’s car, windows rolled down, laughing and making faces at him.

“Hey, Isaac! You’re our hero of all things broken!” the driver shouted as the car raced past him.

“Come on!” Isaac cursed as his car stuttered even more. No wonder James Bond liked his Martini stirred and not shaken, or was it the other way around?! Gosh, this moment was so confusing that he couldn't even remember the sexiest way of ordering a Martini.

Anyhow, the Lotus, winner of so many car races, even the ones Isaac's dad didn't really know about, had come to a complete halt. Now, you could argue—based on the findings of Albert Einstein—that *being at rest* and *any kind of motion* are the same thing to Nature, as the same laws can be observed by anybody, independent of how you move. That however, right now, did not change the annoying fact that Isaac was about to lose this goddam car race.

Just when Isaac thought his current situation couldn't suck anymore, out of nowhere, from the early morning sky, that just an instant ago looked as serene as a yoga video, shot a bright flash of light. Sure, there was more than ample space around Isaac's Lotus. But the flash of light had made up its mind and hit Isaac's dear, holy car with all its might. This produced the second, very honest, "Shit!" to soundwave its way out of Isaac's mouth.

But occasionally, just occasionally, things turn out far better than expected. Imagine a weather forecast predicting more rain after two solid weeks of rain, but instead the next day is sunny all the way. That's exactly what happened to Isaac when he thought his car was wrecked. Instead of falling apart bit by bit, the Lotus started to shake heavily. Isaac could feel the car becoming airborne. Like literally! It sooo gained height, like as high as some of the trees around him.

"Wo-ho, ha, wha…," was all Isaac could put into words.

Whilst he was sorta trying to make a sound, the car, with another bang of course, started shooting forward through the air like a rocket. Instantly, Isaac went from practically being speechless to screaming. He also switched to body language. In this case, holding onto the steering wheel in panic hoping not to shit his pants.

Isaac really had no time to start worrying about what to worry about most, because he thundered by his buddies in the other car. Zip. Just like that, faster than most people can think, he flew by them.

Now it was his buddies who, totally shocked, reflexically hit their breaks. With their mouths wide open, like garage doors, they stared in total disbelief at the freaking Lotus that had just sped by them.

"Since when do broken cars fly?!"

JESSY CLAYTON

AGE: 14

COUNTRY: UNITED KINGDOM

MOTTO: I dare and love to constantly think outside the box! Being happily alive with Nature and taking care of all is what rocks!

IN THE MIDDLE OF GREAT BRITAIN

Around midday, somewhere in the middle of Great Britain, located on the edge of the North Atlantic Ocean, fourteen-year-old Jessy Clayton was jogging through the beautiful forest right behind her parents' house. School had been canceled this week due to renovation work. Jessy's school was about to become more eco-intelligent and climate-smart. All students were proud of this development, as they had pressed consistently for it for well over a year.

Jessy wore her green sweatshirt and a knitted, yellow bobble hat, as the air was brisk this morning. It was the beginning of spring, and the scent from the great pine trees was refreshing. She spotted squirrels, small and large birds, deer, rabbits, and foxes, beautiful butterflies and other numerous insects. Most of the wildlife along her jogging trail was unafraid of her, because not only did she regularly jog through the forest, but she also waved and said "Hello" to them. After constantly doing that for a few years, the animals in the forest considered her a friendly soul and returned her greetings by looking at her with curiosity from a distance instead of instantly taking flight, as if some super destructive maniac had disturbed life in the forest. From her maternal grandparents, who had run a pioneering organic farm, she had learned that all life, plants and wildlife, enjoyed communicating with humans, provided they behaved like reasonable members of the natural network of life.

Often, she thought of revolutionary filmmaker Walt Disney when she saw all the funny, adorable creatures of the forest. As a young boy, even though only for a short period of time, Walt Disney, the famous film producer, had lived with his parents on a farm in the countryside. Later, in his adult life, he always tried to recreate the joyous, wonderful magical feelings, that he had experienced during his interactions with animals and Nature, in his films. This also reminded Jessy of what George Lucas, the legendary creator of Star Wars had said about The Force. That it was "a way of seeing" and "being with life".

For a moment, Jessy stopped running and just inhaled the wonderful cold air with three deep breaths. Then she continued her jogging on the circular route through the countryside. As always, she tried to enjoy what she did, making sure there was a smile on her face. She had read that in the 1990s, neuroscientist Candace Pert had shown that all the cells in the body were continually aware of its overall emotional state, and in addition, in the field of epigenetics, experiments were presented showing that the environment, to a large extent, determined which genes were turned on, and what kind of proteins our body produced. Jessy was totally fascinated by these findings, because scientifically it meant that the ancient insight, that micro- and macrocosm are one, had been confirmed. She always wondered what science-superstar Albert Einstein would have said to that, since he had—using the simple mathematical concept of symmetry as his beautiful guiding star of Nature—unveiled that in the cosmos as a whole everything is one breathtaking harmonious, amazingly balanced and elegant unity. Or what would genius polymath Leonardo da Vinci have said to that had he known? It was said that he had painted his many masterpieces with that unity of all in mind, especially the Mona Lisa with her famous smile. Jessy tried to be aware of that mesmerizing, balanced unity of Nature at all times. But it wasn't always easy with everything that was going on in the world. She was proud to live in a country where there was someone like the Prince of Wales who continuously pointed out the all-interlinking harmony and great unity of Nature, and what humans did to Nature automatically meant that they were doing it to themselves. Naturally, Jessy had joined the *Fridays for Future* movement. It made her feel empowered knowing there were millions and millions of young people of her age also pressing for a lifestyle in harmony with Nature.

Humans, Jessy thought, were a strange species sometimes. Birds, for instance, took the time to enjoy the morning and evening sun as the sunlight was good for their eyes. Why, if birds knew what was good for them, did humans think that running around to constantly set new world records of being stressed out was clever?! There was no first place for having experienced the most fatal heart attack! It's wasn't even an Olympic discipline, or a recognized positive "thing". Yet, it seemed, getting stressed out was the next best thing to proving that one was not yet dead or whatever. And then, there was what she called "Cat Wisdom", or the "Holy Shit Rule".

Once Jessy had observed how a hungry, little stray cat, first relieved herself before rushing over to a bowl with food that Jessy had placed under a tree. This seemed remarkable to Jessy, as there was another, larger cat already feasting on the food, and the possibility that the little cat might not get any at all was real. But to the little cat, it was more important to first relieve herself, before eating. In more abstract terms: the little cat prioritized first detoxifying herself over filling herself up with new proteins (cat food).

"That's discipline," Jessy had thought. You might say, the little cat had first given back to Nature, before taking from Nature again. Even though, normally people don't brag about their shit like they might do with a new sportscar, it was just like peeing and returning nutrients back to Nature. Yeah! In some cultures, many years ago, it was even considered rude to leave a host's house after a great meal without leaving some serious large, nutrient-rich "pile" in the host's garden before saying "Goodbye". That's what Jessy called "Cat Wisdom" or the "Holy Shit Rule". Once you reconnected with the beautiful workings of Nature, it was no longer obvious that a super expensive sportscar (containing numerous unpleasant chemicals and lots of electro smog) was more worth than what you flushed down the toilet.

As always, once Jessy had run half the distance, she sped up. Unfortunately, today she slipped on a wet tree root. Ouch! That hurt!

When Jessy tried to get up, she realized she had hurt her right ankle as well. She was berating herself angrily, as getting home would now be a serious piece of work, but suddenly decided to opt for panicking instead. With her right eye, Jessy had just spotted a long, grey, poisonous viper, with its sigNature zigzag pattern moving down its back. It's not that there are that many dangerous animals roaming around in Great Britain, but that 60-centimeter-long snake was a venomous reptile. Bites were rarely fatal, but the effect was painful nonetheless.

Jessy tried to stay calm, whilst the viper kept slowly moving towards her. Romanticizing Nature did not help at this moment, nor did being totally honest to oneself like, "This is not for me!" or "If this was an ad on the internet, I'd just click it away!" It was a classic case of not having too much fun. It wasn't the absolute worst, but it was a bummer! Being calm and pretending to be invisible also didn't improve the situation, as the goddam snake just kept inching closer, hissing with its black, shiny tongue. Jessy was really in a quandary. She loved Nature, and fought for its preservation, but at this very moment, she seriously wished Nature was as simple as a freshly squeezed fruit and vegetable smoothie. With evermore panic creeping over her, Jessy felt she had to start screaming really, really loudly to scare the viper away. Just as she was about to scream from the top of her lungs, a bright beam of light came shooting through the trees down from heaven and enveloped her. Even though Jessy loved getting mail, this delivery was more than over the top she thought. The brightness of the light and the impact of the strong energy all around her were overwhelming. For a split second, Jessy had the oddest feeling that she was suddenly pain-free and running close to the speed of light.

"Back already?" asked Jessy's mother.

"What?! Um, yeah!" Jessy answered in disbelief. She looked around and found herself standing in the kitchen with, no kidding, a super healthy, green smoothie in her hand.

"How is that possible?" Jessy gasped, staring at the drink.

"You used the juicer like always, dear," her mother replied a bit confused, not understanding why Jessy was so surprised at holding a green drink in her hand.

"But, but…," Jessy spluttered, "but of course!" she finished the sentence. With a smile she took a gulp from the good stuff in her glass. "But what the heck just happened?" she thought to herself.

ABDULAZIZ HUSAIN

AGE: 14

COUNTRY: UNITED ARAB EMIRATS

MOTTO: I dare and love to trust in awesome me and in the powerful talent of others. Happy People and Happy Planet is what

really matters!

SOMEWHERE IN THE DESERT

Abdulaziz stood, gazing around, somewhere in the desert. Sand, rocks, some tiny bushes and a lizard were setting the stage for the real outstanding actor. It was an old building, looking a little bit like a palace. The walls were decorated with the most beautiful ornaments. The leaf-like, round shapes had obviously been inspired by Nature. The periodical patterns, displaying all kinds of symmetric variations, generated a rather peaceful state of mind. This almost caused Abdulaziz to forget his wild heartbeat and the nagging question:

"What had just happened?"

"Who are you?" a little boy of maybe six or seven year of age asked Abdulaziz.

"What?!" Abdulaziz wondered, as he hadn't noticed the boy coming up to him.

"What are you doing here? I've never seen you here before," the little boy again asked the strange teenager who looked pretty confused.

"To be honest," Abdulaziz answered slowly, "I've never seen myself here before either. Because, what I mean, is…" He looked at the little kid, who was maybe half his age, not knowing if he could or should give an honest answer. Because that would sound really weird.

"Are you a serious museum director?" the kid inquired.

"No, why?"

"Because my uncle is one, and he never talks much either, because he's so busy being serious."

"No, I'm not a museum director. I'm just fourteen years old."

"But then you are a hundred percent older than I am, because I'm seven. So, you are very old, because a hundred percent is a lot of age."

"Okay, whiz kid, don't make it sound like I'm from the Stone Age. The last fifteen minutes or so have been confusing enough."

"So, why are you staring at that very old building then?"

"Well, since you are obviously not going to stop being curious. I, I…I'm from Abu Dhabi, and around fifteen minutes ago, I was, very normally, on my way to a store in a shopping mall to return a box of vitamin pills, because they tasted like sunburn, and, …and, I know this sounds weird, but I got hit by a really strong beam of light shooting down from the sky."

"Are you a museum director on drugs? Because, you know, that is forbidden!" the kid interrupted.

"No! No!" Abdulaziz replied, sounding a little desperate. "Look, I told you, I'm not a museum director, and I'm not on drugs or whatever. I'm also not a dinosaur, in case that question suddenly pops up in your mind. Okay! I am, however, on a very strange journey, because, as I said, I got hit by a powerful light. Ever since then, I keep chasing around like a very mindboggled, crazy flash. I end up at these strange places for a few seconds, before…"

WHOOSH! Abdulaziz couldn't finish the sentence as again, with extreme velocity, he shot through the air and away.

The next thing he realized was that he was in another situation he hadn't signed up for voluntarily. He found himself standing behind a podium on the stage of a huge convention center that was packed with hundreds of clever-looking adults. Over the entrance on one side of the auditorium, hung a huge sign:

THE ABU DHABI INTERNATIONAL SUSTAINABILTY
CONFERENCE

"I know this place, this is the hotel's convention center, where…"

"Ah, there you are, Ab!" A man coming onto the stage, greeted him. "Glad you could fill in for your dad on such short notice!"

"What?" Abdulaziz asked the man. It was Professor Salib who was one of his dad's colleagues at the university's environmental science department.

"Your dad phoned me just a moment ago. He said he'd sent you an email to come here and give his lecture as he's held up in a traffic/sandstorm situation. He also sent us his presentation and said, as you know it by heart, you should give it for him."

"Okay," Abdulaziz murmured. He felt it was best to play along and not tell Professor Salib about the interesting "I was hit by a light beam from the sky" incident.

"Great. Ab, are you ready?"

"Sure." Abdulaziz knew his dad's lecture very well, as his dad had it tried out on the family many times. His dad wanted it to be a super presentation, as it was his first talk after the Corona pandemic in front of a live audience again.

"Ladies and Gentlemen," Professor Salib spoke into the microphone, "please give our youngest speaker ever, Abdulaziz Hussain, a warm welcome for his talk on eco-intelligent, positive chemistry."

Abdulaziz was surprised that the energy flash did not catapult him instantly out of the hall. So, once the applause had settled down, he began to talk. First, he described how chemistry was the connecting science between physics and biology, and it was therefore vital, to use chemistry to be inventive and in harmony with Nature. Too often, toxic chemicals were removed from products and replaced with other substances, that were just as detrimental in different ways. This was like replacing stupidity with more stupidity plus idiocy. Then Abdulaziz highlighted that physicists, spearheaded by Albert Einstein, had revealed the importance of symmetry in Nature. Even on the level of quantum mechanics, symmetry or sameness, played the central role, and without it, there would be no interactions between particles, and hence no chemistry and no biology. So, since symmetry also meant balance and harmony, positive chemistry had to ensure that humans acted in harmony with Nature.

All Abdulaziz could remember was the great wave of applause as he said this. Then all became blurry again, as the energy flash fast-tracked him back to the old building with the beautiful ancient ornaments.

"I don't think you are a museum director," the young boy said to Abdulaziz.

"I'm so relieved," replied Abdulaziz.

Since the kid did not ask anymore impossible questions, Abdulaziz kept quiet, too, and squinted in the sunlight, at the beautiful patterns in front of him. It caused him to think of the internet, how this technological revolution had connected billions of young people who, together, fought for their future, in a pattern of unity around the globe. This technological innovation had reconnected his generation with Nature, with the need to preserve and regenerate the premise of life. In that moment, he felt how the little kid grabbed his hand and silently held onto it. They both knew things had to change on this planet, fast.

JAFARI KAMAL

AGE: 14

COUNTRY: TANZANIA

MOTTO: I dare and love to be super unique and forever bold. WE all have to feel GREAT, think BIG and not fit into a lifeless, tiny, standardized, dead mold.

SERENGETI NATIONAL PARK

About the same time as Abdulaziz was giving his lecture at the "Abu Dhabi International Sustainability Conference", something most remarkable with a five star "WOW" factor had happened in a small but modern school near the new Serengeti Network Ecosystem Research Institute. The thing, actually just one of the things that were so stunning, was that the entire school was now covered with something absolutely cool: numbers.

But let's go back a little in time to the very morning of this unusually special day. Like all mornings, people wake up. That's an international standard that works pretty well everywhere. That morning, 14-year-old Jafari woke up and felt super energized. He felt it right away: This was a game-on kind of day. As always, Jafari looked outside his window to take in the beautiful sunrise over the Serengeti National Park. Today, there was a very inspiring, magical glimmer across the park. It was as sparkling as the first day Jafari had moved here with his parents and one of his younger sisters, named Keisha.

For over two years before that, Jafari had constantly nagged his dad about how the family should do something about preserving the environment and by doing so, the future of human civilization. His dad was an IT specialist at a big international logistics company in Cape Town, South Africa. Jafari had kept pushing his parents relentlessly on that subject, helped by his little sister Keisha and his much older sister Leal. Eventually his dad had talked about this with his boss at work. Now, the boss turned out to be a real cool guy and went to his boss, and since the company was already sponsoring a marine biologist in Cape Town, the bosses of all bosses, that is the Company's Board, decided to fund a new research institute in the Serengeti National Park. Jafari's dad became the Institute's director, and he and his team built artificial intelligence systems to gather data in order to improve the balance between Nature and human society. That's how the whole family came to move to the Serengeti National Park in Tanzania. Except for Leal, who studied at the African Institute for Mathematical Sciences, which was part of the *Next Einstein Initiative*, at a facility in Tanzania.

Jafari's first cool thing that morning was saying goodbye to Leal and Keisha. They were both travelling to an international conference in Stockholm to attend the signing of the Global Green Deal. Both sisters were part of a global delegation of young people that had been invited to take part in the historic event and also sign the Treaty. Naturally, there was a lot of hugging, more hugging and even more hugging and kissing going on. Even the family's colorful, small parrot, a so-called Fischer's lovebird, who had shown up one day and decided that he was part of the gang, was going crazy. That bird always wanted to be part of the action, and contributed as much as possible to the ever-present fun in Jafari's family, which was always very loud and very direct.

Eventually, as the girls had a plane to catch, saying goodbye came to an end. As he was late for school, Jafari jumped on his bike and rode the two miles to the new school that had been built next to the institute. Normally, this was a routine bike ride. But today, it turned out to be so out of the ordinary, that even if you were a person always expecting the unexpected, you'd be stunned. What happened that morning, while Jafari was riding his bike, was more than just spectacular. It was literally the whole Serengeti package!

Jafari was peddling along, when all of a sudden, all the different types of animals that lived in the park showed up around him, even the large crocodiles that were known to be a bit rough around the edges. It was simply overwhelming, really loud and very dusty, and the earth vibrated like crazy: lions, zebras, giraffes, those super heavy hippopotamuses, leopards, cheetahs (which you cannot outrun on a bike), gazillions of wildebeest, countless birds, snakes, beetles, elephants. Instantly, Jafari opted for an all-out panic procedure, which meant he tried to ride his bike as fast as possible towards his school. Even people who have a lot of friends would not have known how to handle that much attention.

Jafari somehow managed to reach and safely rush into the school building, whilst thousands and thousands of animals gathered around the school. The students and teachers were scared, as in like really scared. Because the animals just stood there. Quietly they all stared inside the school building through the windows. Quietly and peacefully, like lions next to zebras, or funny-looking monkeys sitting comfortably in a crocodile's open mouth or huge bumble bees kicking back beside a super large spider on a Wild Stockrose that adorned the window ledge; the sort of thing that you only see happening in a movie. Even all the colorful fish, seahorses, crayfish and water snails in the classroom aquarium stopped swimming around and stared at the students. It was a real bonding experience, and after a while even the humans inside the school building chillaxed. The situation was almost poetic!

Since the animals didn't move, apparently waiting for the next big thing to happen, the teachers and students decided to proceed with their classes. It made more sense than say, trying to reinvent the wheel or something like that because nothing, seriously nothing, no gun fire, no rangers in helicopters, no honking cars, nothing encouraged the animals to do anything but wait and stare into the class rooms.

It was a long school day. At some point, Jafari dozed off a bit when his teacher was talking about the golden ratio and the number Phi. This number was irrational, meaning it goes on forever into infinity. Loving numbers and all things relating to artificial intelligence, the notion of infinity caused Jafari's mind to drift off and think about computers.

The father of modern computers, Alan Turing, like other mathematicians or physicists such as Roger Penrose, had pointed out that not everything can be computed. But the things that were not computable could still be regarded as being mathematical. Galileo had said that the language of Nature is mathematical, and science, as mathematician Henri Poincaré had pointed out, needs mathematical infinity (which is not something you can ever pin down or calculate with a concrete number) to work. Why? Because a scientifically established law of Nature is only a GENERAL law, which is true everywhere, if it is valid for infinitely many cases. So, the law of gravity, for instance, is *general* as it is valid with respect to a red, juicy apple, a love letter, an elephant, even a booger from your nose or all the other countless, massive objects that we can see on planet Earth. The law of gravity is true regarding beautiful planet Earth itself, regarding the enormously bright shining sun, or all the other unique planets of the solar system. Gravity is valid with respect to all giant nebulae, rocks, planets and stars in our galaxy and regarding all other radiating and glittering galaxies in the vastness of the entire universe. So, the mathematical concept of infinity is part of the magical reality we all live in to make GENERAL statements or reliable laws of Nature possible. Nature needs forever mysterious, endless infinity to work. Infinity is part of the universe's large scale, that is ruled by the majestic, stern, smooth law of gravity, and it is present on the universe's small scale, governed by fancy-pants, quirky-looking, never-resting quantum mechanics. The quantum state of the universe, for instance, is described by physicists as infinitely many possibilities that are being tried out at the same time. The universe as a whole is, indeed, a magnificent, super cool simple wonder! Because it is so clever as its simple unity or fascinating, all-encompassing wholeness allows the intangible mathematical concept of open-end infinity to be a constructive part in its creative being on all levels.

Bang! That's science at its best! Because, as Albert Einstein had also pointed out, the mysterious is the source of all great science and art. It makes people wonder, stand and gaze at the marvelous cosmos with awe. And so, it was clear to Jafari that even Artificial *General* Intelligence, or so-called strong AI, therefore needed to have a meaningful, coherent grasp of the important role of infinity in Nature in order to be GENERAL, to be able to fully align itself with Nature's simple wholeness, that is all-encompassing as it includes infinity.

Jafari always felt that Einstein's true legacy was his work regarding the core role of symmetry in Nature. It was the concept of balance or harmony which enabled Nature to make infinity a positive part of the whole. Unfortunately, humankind had made a big choice by destroying Nature's harmony—not a wise choice as his generation knew, and which scared them and made them lie awake many nights. But it also united them around the globe, to fight for their future.

Jafari awoke from his daydreaming and could hear his teacher finish his explanations of the number Phi: "…and that's why Phi is called irrational, because it goes on forever."

"Like school," Jafari mumbled.

He said that a bit louder than he intended. Everybody heard it, and laughed, including the animals, judging by the funny sounds they made.

"Well then," the teacher said, "as you're such a crowd pleaser this afternoon Jafari, why don't you entertain us with something really impressive. Like writing down as many digits of Phi as you can." With a smile the teacher handed Jafari the red chalk and pointed at the chalk board.

At that very instant, a white, very bright, strong beam of light shot down from the sky, into the classroom and hit Jafari. The next thing Jafari noticed was that he was writing countless digits of Phi all over the chalkboard, the classroom, the tables, chairs, the other classrooms and then all over the school. It all happened so quickly that it was done before Jafari could fully realize it.

His classmates and teacher stared at the myriad of digits all around them in disbelief. The chalk in Jafari's hand was glowing and smoking, a testament to how fast he had written the numbers. The only ones who seemed to be super 'okay' with what had just happened were the animals. Peacefully and contentedly, they all turned around and walked, like happy fans who had just witnessed their favorite basketball team win, back home into the wilderness of the beautiful Serengeti.

MEI CHANG

AGE: 14

COUNTRY: CHINA

MOTTO: I dare and love to so believe in the incredible me, my sexy ability to create both my own cool destiny and our eco-intelligent future that I call prosperous, amazing WE!

BEIJING'S DONGCHENG DISTRICT

"Scream! I'm gonna scream!" muttered Mei Chang to her 14-year-old self in utter frustration. But instead of actually screaming, she just turned on some hot Chinese pop and rap music. She liked female singers like Jolin Tsai, G.E.M and Diamond Chang. Mei sat down on the comfortable pink couch in her room and glowered at the object of her frustration. It was that damned spaceship which she had designed for a school project. The right wing of the spaceship just wouldn't stay attached. It fell off every time, no matter what kind of glue she used. This was insane and totally unreal.

"I should have gone with that first round, colorful spaceship image that had popped up in my mind when I thought of designing a spaceship. But no, I had to make it more difficult!"

Mei turned up the volume of the music. She loved design and engineering. Her mom was a fashion designer, producing eco-intelligent garments, and her dad was an engineer building eco-intelligent, climate-smart cities in China. Actually, Mei considered everything to be a design event. Her grandmother, a revered doctor of Chinese medicine, had introduced her to the concept of Yin and Yang, the Chinese Five Elements, as well as Feng Shui, at an early age. Granny had always pointed out that Yin and Yang were one and to be found everywhere, like in the Forbidden City that was part of her neighborhood. According to Mei's sweet Granny, the universe was a self-similar, self-organizing wonder. The old sages, Granny had told her, had discovered that marvelous flow of unity in Nature, the energy called Chi, and had described its constructive, cyclic phases with Yin and Yang, the Five Elements and the animals of Feng Shui.

It was amazing, but this very omnipresent harmony of Nature, mathematical physicist Albert Einstein had unveiled with his strong imagination and had underpinned these groundbreaking, revolutionary insights with scientific facts, just like the great mathematician Emmy Noether had done afterwards as well. The mindboggling simplicity and elegance that Einstein had revealed carried the mathematical name of symmetry. Actually, it was a so-called **LOCAL** SYMMETRY that was just cool and awesome beyond words. The word "local", of course, denoted a concise area where the symmetry did its thing of uniting seemingly different phenomena.

The first local symmetry leading up to the second, mega, kick-ass local symmetry was that each massive object, small or extra-large, had two identical or symmetric masses. That's already pretty interesting.

Based on these two identical masses Einstein went on, thinking the implications of that local symmetry of the two masses through in his powerful mind, that to Nature both *being at rest* (not moving your ass) and *acceleration* (enjoying ever more speed) were locally the same, identical, or in scientific vernacular: symmetric. The funky conclusion eventually was that *being at rest* and *any kind of motion* are the same to Nature, and therefore, the laws of Nature were the same, that is symmetric, everywhere at all times. That's a very original design: Unity, harmony, balance. That was some real good, effective and efficient thinking. Because if you keep stuff simple, it can really get big, like enormously. And the cosmos had a pretty vast presence as scientists knew.

But Mei had also liked the following: Since *being at rest* and *acceleration* or *even any kind of motion* are the same, a unity to Nature, this meant that this unity contained infinity. Why? Simply because one could imagine infinitely many kinds of motion—like fast, faster, insanely fast, crazy fast, even faster again and so on; even if one respects the upper limit of the constant speed of light in empty space. So, the super cool, kick-ass local symmetry that Einstein had discovered contained infinity. As this symmetry then applied to infinitely many scenarios, it was kinda logical, Mei reasoned again, that this local symmetry was true everywhere. And as that meant that Nature only saw itself as a breathtakingly beautiful symmetric unity, the laws of Nature had to be the same everywhere at all times for all observers. Now that's a real nice touch of both artistic and mathematical cleverness and sexiness. No wonder, the universe had been in style for more than 13,8 billion years or so.

This awesomeness really gets into your head. In Mei's case it caused her to a) buy organic produce from local farmers and to b) pay attention to her health and look colorful and confident, like the cosmos. Accordingly, she liked fun fashion, big, shiny attachable eyelashes and colorful fingernails. Afterall, if Nature enjoyed being creative in its time and space with independent simple elegance, that could rely on the forever mysterious infinity term as countless possibilities, so could she.

Yet, the mega annoying question then was: Why did humanity behave like a crazy assistant with an opinion? Or put differently, why did humanity meddle with Nature's successful concept? It's one thing to want to leave a dent in the universe by becoming the world heavyweight boxing champion or something, but to outright destroy the omnipresent concept of harmony, balance or symmetry?! It seemed such a waste, literally. Why bet against reality if you didn't like losing?

Strangely enough, there was hope around the corner. Artificial Intelligence equipped with sensors could actually and most importantly locally enhance the symmetry of products by allowing them to be in greater harmony with their surrounding: Like autonomous cars that could better align themselves with their environment, additionally producing more beneficial effects. Such as fleets of autonomous taxis dramatically reducing the number of cars needed on the roads, allowing humans to return parking spaces back to Nature. Nature could then work her regenerative magic and optimize the symmetric network first locally, and then globally, so improving the environment again.

The same was true in respect to food production: Technologies like precision fermentation or bioreactors could locally produce food of much greater quality and usefulness. This would not only end the cruel treatment of animals, but it would again also free up vast amounts of land that could be returned to Nature and her precious wonderful circular workings. Authors like Tony Seba and James Arbib even spoke of the possibility of an "Age of Freedom" for everyone in the near future.

Much, obviously depended, as Mei reasoned, on whether

scientists, companies, consumers and policymakers, understood that technology needed to have a clear purpose, as political economist and sustainability scientist, Maja Göpel had highlighted. That could only mean one thing, in Mei's mind, respecting and promoting the core principle of Nature that Albert Einstein had discovered, symmetry. It also included making sure societal needs were met. Chinese technology investor Kai-Fu Lee had pointed this out. Making sure that the wealth created by artificial intelligence technology enabled humans to become even more human, to experience not less love, but more, much more.

Mei got up from her couch, threw the unnecessarily complex spaceship on the floor, and with her super sexy high-heel light blue pumps she crushed it into a thousand pieces. Right after she had changed the status of the spaceship from being an annoying contraption to, "Go to hell, sucker!" her dad entered her room.

"Is everything OK, Mei?"

"Sure. As the commanding captain of my spaceship, I just decided to allow myself to be upset, to retire that stupid machine and to make things simpler and more beautiful."

"Sounds like a trendsetting, promising design plan," her dad said with a smile.

He knew by the look on the face of his talented 14-year-old daughter, that he had no choice but to agree with her in order to make a quick, graceful exit with a note in his head to not worry so much about Mei's sweet, but also unpredictably innovative mindset. Returning his smile, Mei showed "two thumbs up" to encourage her dad to leave the room even faster.

She sat down at her desk and drew the spaceship in the way she had actually wanted to design it in the beginning: Round! In that very moment it was Mei's turn to get hit by the bright beam of white light. It shot right through her window.

The next thing she noticed was that she was flying inside a round-shaped spaceship through Beijing, close to the speed of light. Now she did scream for various reasons:

1) Whoa! Highrise right in front of me!
2) WHAT IS HAPPENING?!?!
3) Wow, almost hit the top of that building!
4) What's that tree doing in front of me?!
5) I lost an eyelash!
6) P-A-N-I-C

Before Mei could fully realize what was going on, the spaceship made a loud BANG. Then she was crushed breathlessly into her comfortable pink couch, holding a round, colorful spaceship in her hands. From outside the door, she could hear her father's somewhat worried voice:

"Are you OK?"

"Never better!" she replied, wondering if she had just been dreaming, or if not, how had she signed up for this?!

JAFARI'S HOUSE

It was way past midnight. But Jafari couldn't sleep. He rolled from side to side in his bed, as the events from the afternoon were keeping him awake. Nobody could explain what had happened. Countless pictures of the animals staring into the classroom were now all over the internet. That generated a lot of buzz. Yet, the picture of the school being covered with countless digits of the number Phi got even more attention. How, just how had he done that?

Finally, Jafari turned on the light and got up. He turned his laptop on and sat down at his desk on the chair in his yellow T-shirt and blue shorts. He yawned. That was a good thing. For a moment he looked out of the window of the wooden lodge he lived in with his family. A full moon shone brightly onto the vast wilderness. Right now, he could only hear grasshoppers chirping.

"What's that?" he asked himself. On the screen of his laptop a computer game had just appeared, one that Jafari was pretty sure he didn't know. The action happened via a colorful, round spaceship, called System 1, that had to shoot down dark drones.

It wasn't the most challenging game. Bam. Bam. Another point. Bam, more points. Bam. Bam. But it was OK with respect to being half asleep and not knowing how to sleep and just wanting to kill some time… The game quietened down his mind and put him into an almost timeless state. He was about to doze off.

Suddenly, in the middle of the bam-bam action, a popup window appeared with a ring tone. The message read:

YOU HAVE FOUR NEW FRIEND SUGGESTIONS ON STELLAR PLUS NETWORK

Jafari clicked on the button next to the message and found himself on a social media and networking website. He didn't know the website, and certainly didn't want to join it. A strange energy, however, guided his hands. He didn't know anymore, if he was awake or not, he stopped resisting and typed in his password. Access was granted.

A bright flash accompanied with a loud WHOOOSH sound jolted Jafari and totally woke him up again. In disbelief he stared at the screen, where he saw the faces of the four friend suggestions.

"What the ??!" He began to read their profiles. "Isaac Al, 14, United States, Physics professor. Come on! A physics professor at age 14?"

Jafari looked at the next profile. "Jessy Clayton, 14, United Kingdom, Biology professor. What? Is this a joke?"

Next profile. "Abdulaziz Hussain, 14, United Arab Emirates. Of course, a Chemistry professor. What kinda website is this? For dumb teenagers?"

Shaking his head, Jafari looked at the last friend suggestions: "You, MissLady, are Mei Chang, 14, from China and, how sweet, a Design- and Engineering professor. Yeah, right. And I am…,"

Jafari looked at the data stated under his photo and gasped: "Artificial Intelligence and Mathematics professor?! Not funny. Who's pulling this shit on me?! I don't like practical pranks. Come on. Who are you?!"

With another bright and loud flash that made Jafari go "HA!" and move backwards real fast, another pop-up window appeared underneath the four friend suggestions. It contained a quote by Albert Einstein.

"Oh, that looks real mature," Jafari mumbled as he started to read the quote out loud:

"Based on our experiences so far, we have reason to be confident that Nature is the realization of the simplest conceivable mathematical concept."

Albert Einstein

(On the Method of Theoretical Physics, June, 10, 1933, Author translation)

Jafari looked at the quote in disbelief looking right into his webcam. Then he grumbled:

"Could you, whoever this is, stop suggesting things. It's way past midnight, and I don't wanna meet no new friends online and, and…"

A new popup window appeared. This time displaying a quote by two other physicists:

"There are (...) enough pieces of the jigsaw puzzle in hand to know that symmetry is fundamental to all of it. The abstract concept of symmetry and its relationship to the physical world is enduring and here to stay."
Leon M. Lederman & Christopher T. Hill
(Symmetry and the Beautiful Universe, 2004, pp.20-21)

"Hey!" Jafari protested. "What did I just say about not suggesting things. Okay, that's it! I'm just gonna turn the laptop off right now. And tomorrow I'm gonna search the entire world wide web to find out who did that, and I will not come in peace! Repeat: I will not come in peace! I'll find you and open a serious can of whoop-ass on you!"

But before he could turn off the laptop, the four pictures of teenage professors started to radiate in a bright light. A title card read:

THANKS FOR ACCEPTING THE FRIEND REQUESTS

"I sooo did not!" Jafari objected. "This is…"
A new message appeared:

CONGRATUALTIONS

PROFESSOR JAFARI KAMAL!

YOU JUST WON A TRIP TO THE ETERNAL CITY

This was followed by a super white flash that burst out of his laptop's screen. The entire room was filled with the powerful light.

CHAPTER 3

For a moment, Jafari was stunned senseless. The bright light, swirling around like a super powerful whirlwind was just too much information for him, so his brain and body reduced its functions to vital subprotocols, i.e., only keeping up the core bodily functions.

A split second later, Jafari woke up again. He blinked and tried to focus. Apparently, he was lying on the floor, and it seemed like that four faces were staring at him.

"Welcome, Professor Kamal," Jessy Clayton said to Jafari.

"What?!" Jafari could now keep his eyes open and focus. "Again, what?!"

Totally confused he looked around. The room was a round shape, part of the wall was equipped with blinking electronic devices, screens and signals. Another part of the wall was a huge window behind a control panel with chairs. It looked like some sort of spaceship interior.

Isaac grabbed Jafari on his arm and pulled him to his feet.

"Good to have the world's leading AI expert on board."

Jafari still looked at everything in disbelief. The four teenagers in front of him were dressed in some kind of spacesuit uniform, while he was still wearing his T-shirt and shorts.

"Okay, something is not quite OK. I know you dudes! You're the, the guys from the friend request thing…I, I can't remember how I got here. How, how did you people get here?"

"We don't know," replied Abdulaziz.

"That is not a satisfying explanation," Jafari protested. "What is this round thing here anyway? It looks like too many Christmas decorations in a relatively small space configuration?"

"This," Mei responded, "is my spaceship, called System 1. In other words, I'm the captain of this vessel."

"You?! But you're, like just like a 14-year-old Chinese girl!" Jafari stuttered.

"What's your point, Jafari?" Mei inquired swiftly.

"I, I don't have a point. All I have is a lot of question marks, and those are not the hilarious ones. No, my question marks are pretty confused and a bit pissed at this whole…situation."

"I think it's best to calm yourself." Jessy smiled at Jafari.

"Calm?! Calm myself? About what? We're not even having a real conversation. Because we talk about things without even talking about them, because we don't know what we are talking about." Jafari sighed. "This is like when politicians talk. I am definitely not up for this. I'm leaving. Right now. Because this can only be a messed-up dream.”

Jafari turned around with the clear objective of wanting to walk through the spaceships wall.

“Ouch!” Jafari seriously hit his nose when he made contact with the wall.

“*Sorry, Professor Kamal,*” an electronic, female voice could be heard, “*these walls are not made for walking through.*”

The four teenagers had to giggle, while Jafari rubbed his nose.

“Who’s that voice, that apparently thinks comedy is the answer to everything.”

“That’s Jacky, the voice of the spaceships strong AI or AGI, that is Artificial General Intelligence. I can tell, the two of you are off to a great start together.” Mei winked at Jafari.

“You think so, huh?!” Jafari moved his nose around, checking that it was not broken or something worse.

“You’re good at this thing,” Jessy laughed.

“What thing, Jess?!”

“Visual comedy. The thing you do with your nose.”

“That’s not a thing. Besides, why are you all dressed like that? Where did you all get these champion-style costumes, and more importantly, why don’t I have one?!”

Mei grabbed Jafari by the hand and guided him to a door in the wall. The door opened automatically. Inside the closet hung two more uniforms.

"We don't like to think of this as a costume. More like a uniform. This one should fit you."

Mei handed one uniform to Jafari, who put it on immediately.

"Who is the last costume/uniform for?"

"Santa Clause, your mom, the U.S. president…"

"You have no idea. Great!" Jafari said as he finished putting on the spacesuit.

"Now, what? Fly off to Rio for the Carnival?"

"We don't know. We were kinda hoping, you might have a clue," Abdulaziz said.

"Well, before I appeared here. My laptop told me that I had won a trip to the Eternal City."

"*AKA Rome,*" Jacky's voice could be heard, "*also known as the capital of the world. I will set course for Rome.*"

System1 flew a slight curve to the right. Mei checked the screen of the control panel. "We are heading north, across the African continent towards Rome."

"Or," Jafari interrupted, "instead of flying to Rome, why don't you fly me back home!"

Just as he had said these deep, meaningful words, a loud alarm went off. This was followed by a mighty BOOM that heavily shook the spaceship.

"Now what?" Abdulaziz yelled.

"I hope this is the teaser to the spaceship's haphazard onboard entertainment program!" Isaac added, holding onto the wall as another hefty BANG made everything tremble.

Mei looked at the radar and with shock saw many other, rather small flying objects closing in and firing at them. "Nope, it's way more unsettling. It looks like we got bad company."

Jafari looked out of one of the small windows and spotted numerous drones approaching and shooting laser cannons at them.

"There are countless drones. They don't look like they wanna invite us to a seminar about best practices either."

"Buckle up, guys!" Isaac screamed.

The five heroes jumped into their chairs, sat down and fastened their seatbelts. More cannons hit the spaceship.

"If they wanna say `Hello!´ so badly, they should just follow us on social media." Jessy shouted.

"Shields are down to 60 percent!" Jacky, the onboard AI reported.

"So, now would be a good time to get rid of them!" Jafari yelled.
They were hit again, and the spaceship entered into a downward tailspin. They all screamed at the top of their lungs.

"I'm switching to manual!" Mei announced and grabbed the joystick.

"Whatever keeps us from exploding!" was Isaac's motivational comment.

"That's the spirit!" Jafari and Jessy added simultaneously.

Mei started flying wild maneuvers to escape the attacks.

"And I always dreamt that if I was in a spaceship, it would be because some friendly, charming and vegan, price-charming alien wanted to tell me, 'You are not alone.'" Jessy sighed.

"Right now, I kinda wish we were alone," Isaac tried to joke.

Mei made System 1 fly ever more wild maneuvers: looping, zigzagging, chasing through small, rocky canyons. She even initiated the camouflage mode and changed the beautiful bright colors of her spaceship to black along with broadcasting disruptive signals. But the high-tech drones didn't care very much. They just kept firing even more.

"Damned! These nasty bitches aren't even fooled by the camouflage mode."

"The chemistry between us and these overreacting flying contraptions is not good!" Abdulaziz added with a trembling voice.

One of the drones managed to fly close by. When Jessy noticed the machine, she saw the label written on it its side said:

SUBTRACTOFORCE

"Subtractoforce. Does anyone know that name?"

"No, and I sure don't want that word on my next birthday cake!" Abdulaziz yelled.

"I've run several simulations and analyses in quantum parallel," Jacky announced. *"The drones seem to be having problems flying continuous spirals and have a very negative vibe!"*

"Meaning what, Jacky?!?" Mei shouted.

"I'd suggest keep flying spirals while advancing forward at the highest possible speed. Plus, to counter the bad, shit-crazy swarm intelligence chasing our asses, some funky rap music!"

Jacky took over the steering of the vessel while she started some real loud rap music, blasting it into the night sky via external loudspeakers as well.

"OMG!" Jessy screamed as the spiraling flight maneuver made it impossible to tell with certainty where brain or stomach were at any one moment.

"I hate these barf-bag-friendly situations!" Abdulaziz confirmed.

Mei looked desperate for another technical fact: "Our shields are down to 40 percent. I deactivated the camouflage mode to save energy."

Jafari started working on the computer's touchscreen as best as he could.

"What are you doing?!"

"Trying to free up energy for the shields. Like by making your algorithms more efficient or rerouting energy from your oversized, onboard makeup compartment!" Jafari replied.

"How do you know I have a rather large makeup compartment on this spaceship?!"

"I didn't, was just a super wild guess, given your good looks."

"Are you hitting on me!?"

"No, that's what the drones are doing."

They were now approaching the three great pyramids of Giza, as their escape attempts had pulled them off course. The wild combination of flying spirals accompanied by countless laser torpedoes, many of which overlapped with each other to cause even brighter rays of light, and the loud rap music created a strong reaction in possibility space.

Possibility space, as scientists had realized, is a purely mathematical, omnipresent, super extremely tiny (to the point of physical-nonexistence) space of many infinitely convoluted dimensions that, in a timeless fashion, contains all kinds of ideas, data, designs, concepts—in short infinitely many possibilities or alternatives. Even Nature's evolution, as mathematical biologist Andreas Wagner and his interdisciplinary, international team had discovered in a groundbreaking fashion with the help of AI, relies on this higher-dimensional space to get some serious traction into evolutionary processes.

The point being, the super turbulent orchestration of spirals, laser torpedoes and funky "Yo, bro!" rap music had produced some stormy waves in possibility space, just above the head of the Sphinx in front of the mighty pyramids. This led to an untamed culmination of a lot of appealing possibilities and interesting options, blasting right into the monumental statue. The result was, well, you could say that the otherwise poised Sphinx got a bit rowdy and undisciplined. Certainly not in keeping with her world-famous brand of being unmoved (for thousands of years), the Sphinx suddenly became ferociously alive. In a gaga-kinda way, the huge animal-and-human-like beast started lashing out against the recklessly shooting drones as if they were annoying, super self-possessed mosquitoes. The drones were totally caught off guard by this unconventional version of a cat-woman, which gave the drones a lesson in "Since you don't know what anger management is, I'm just putting you out of business!" In other words, the overpowering force of the Sphinx's creative moment made the drones, one by one, crash into the desert's sand.

Luckily, from the point of view of our five heroes in their spaceship System 1, the whole unconventional eruption of the Sphinx out of her deep, beautiful meditation had an exciting, life-saving touch to it. Because System 1 had been able to shoot back up into the night sky, right before the Sphinx had started her knock-out fest with the drones.

Suddenly, life felt more easy-going again in System 1, and the five heroes expressed their joy by doing a little dancing along with the groovy beat of the rap song. Isaac even impressed his team with a little, selfless show of breakdancing.

Down in the desert, the Sphinx resumed her original, self-assured posture with her heavenly, Mona-Lisa-like smile. In her cool, placid way of timeless, radiating self-confidence, the monument appeared more than ever as a symbol of Nature's unity and wholeness.

CHAPTER 4

System 1 could finally return to its course towards the Eternal City, the Capital of the World: beautiful, inspiring Rome. This city and the entire country of Italy served the greatest minds, artists, polymaths, scientist and philosophers as a source of uplifting creativity and a place of finding wisdom and oneself. That is what had happened to the great German poet and polymath Johann Wolfgang Goethe in the late 18th century. His famous trip to Italy was a life-changing experience. It had been a travel enthusiast's dream for him, so to speak. Seeing the mesmerizing beauty of the countryside, diving into the rich, timeless cultural heritage of Rome and understanding the unity of all, the wholeness of Nature, while spending three months on the island of Sicily had lit an eternal flame of happiness inside the great scholar.

Inside the spaceship, that was now flying across the Mediterranean Sea, the crew was still wondering about what had just happened. No one could find the right words for quite some time. Finally, Jafari ventured a comment far outside the traditional concept of reality.

"I know my fear of invisible snakes, for instance, is odd, but this dynamic badass situation was just totally crazy. Ancient monuments are not supposed to move, at all, ever. That's just not normal."

"I think the word `normal´ is definitely no longer part of our reality," Isaac replied.

"How is that insight helping, Isaac?" Jessy said.

"I don't know. I'm just saying," Isaac explained, "that, mathematically, unlikely things start happening when the path of normality becomes ever smaller. Then the most unlikely events get more probable in relation to what used to be a wide range of stable normality."

"Okay, mathematically, I'm totally with you, Isaac. But what's causing this crazy stuff?! It can't be my usual midnight appetite for muffins, because all this zig-zag flying is making me feel queasy."

"Climate change. Pollution. Normalcy is being attacked from infinite angles these days," Jessy reasoned.

"Then let's hope that Global Green Deal will be successful," Abdulaziz added.

"We are now reaching the coast of Italy. We'll be approaching Rome shortly," Mei announced looking at the map of the computer's screen. "Anything you want to add, Jacky?"

"I cross-referenced your backgrounds. This generated clusters of overlapping data. Most outstanding were two sets of data: 1) your knowledge of the current environmental crisis and 2) your awareness of Einstein's true legacy, which is his insights about the core role of symmetry in Nature. This is very unique, because most people are nowadays aware of the climate crises, but not of the groundbreaking insights by Albert Einstein about the core principle of Nature. Very odd, indeed, because from an Artificial Intelligence point of view it would be much more reasonable if everyone knew about the most important idea that underpins our reality. In that context, I just want to mention that Rome was founded by twins; highlighting symmetry, harmony, balance, which was also the central theme of Romulus' reign—for good and for bad—which at the end of his life was turned into a transcendent concept in the form of the deity Quirinus. The name Quirinus itself is also seen as a replacement for the god Janus."

"The double-faced god," Mei said quietly and continued, "like Yin and Yang."

"We've got that idea of binary oneness depicted in ancient African artefacts as well," Jafari added nodding affirmatively.

"Maybe it should not surprise us after all," Isaac said. "I mean, we live on a planet that has two symmetric, balanced, identical masses. Einstein used this fact to start out on his groundbreaking journey of discovering that symmetry, which at first was his guiding star, eventually turned out to be the core principle of the cosmos. Kabbalists also see the universe in a symmetric and unified fashion: one sphere resembles all the others in the tree of life that depicts the marvelous entirety of the cosmos."

"With the Greek word `cosmos´ meaning order," Abdulaziz said. "Not only are we humans built in a symmetric way, the chemical element Carbon, the root cause for global warming, is essential to all organic matter via its many compounds and depicts symmetry or balance in an outstanding fashion, too. Carbon is not too small, not too big, and it's both stable and eager to react."

"And the Prince of Wales, wrote an entire book about how harmony is the core premise of Nature," Jessy added.

"That's what I love about being AI," Jacky said. *"You suck in all that data, connect it, add all up: Bam! You find these hidden realities that are not instantly accessible to the human senses."*

"No wonder conversational AI is all the rage right now. Who would have thought that AI can bring culture to our group?"

"Why not!" Jacky commented, *"you just have to know the name of the game to respect that reference frame."*

"Hey everyone, why not take the rest of the time to unplug a little more, as who knows what to expect in Rome?!" Mei said.

The five of them returned to being quiet and trying to relax. They could still feel an enormous amount of adrenalin rushing through their blood. Their bodily systems had not yet calmed down, and they were all still trying to come to terms with the extraordinary challenges that had occurred, and not to mention the mystery of them.

Jessy took out an iPod.

"What are you listening to, Jessy?" Abdulaziz asked.

"Whale sounds. I like animals who have a strong sense of community. Want to check it out, Ab?"

"No, I'm good."

Before Jessy could start listening to her whale sounds, Mei abruptly turned around in her chair and said with a loud, almost shocked voice:

"I know what's wrong?!"

"And I have a feeling you wanna share that insight."

"You bet, Isaac. There are only two girls here. Two, just two!"

"You mean like!?" Jessy asked.

"That's right girlfriend. The unescapable fact is that we are underrepresented here. Like organic produce in supermarkets. Which is so annoying."

"You're right, Mei. If you think about it, in untouched Nature, you only got organic stuff. That should tell you something."

Jafari joined the unfolding conversation: "Sooo, what are we talking about? You being underrepresented, or the lack of supermarkets and industrially processed pink chunk food in the jungle?"

Mei and Jessy gave Jafari an angry look.

"Hey, Jafari, dude, keep it simple," Isaac whispered to Jafari. "Keep us guys out of this conversation."

"Too late for that," Mei countered. "You're part of this, guys. 'Cause you are three, and we ladies just two."

Abdulaziz tried to appease the situation, "But statistically, you women live longer."

"But you get still paid more, statistically!" Jessy objected.

Again, Isaac tried to convince his buddies to stay out of this discussion. "Guys, we can't win this conversation!" Then he turned to Mei and Jessy and in a very nonchalant way said, "But you, ladies, win. Big, big time."

Jessy and Mei hi-fived each other.

"But they are still just two," Abdulaziz noted.

Isaac sighed: "Man, let me introduce you fellows to the concept of SHHH!"

Mei agreed with Abdulaziz: "He's right. We still have to do something really major about this injustice."

"Ladies, ladies," Isaac said, trying to refocus the group. "How about, we postpone this discussion to a later, brilliant and historic moment and change the subject to 'us arriving in Rome.' Because that is what's happening right now, and Rome, oh Rome, I am sure, wants to see you happy."

"Okay. We'll talk it out later," Mei said. "But only because Rome is such a beautiful city."

Approaching from the south, the spaceship flew over Rome along the river Tiber. It was late afternoon. But no one really paid attention, as reality had become way more creative than ever. The color of the spaceship changed to black, so that it could hardly be seen anymore.

The city looked spectacular in the sunset as the metropolis' lights had already been turned on. All the major sightseeing highlights were pointed out on the spaceship's big front window via augmented reality.

Apart from producing expensive, high-tech, live-action spectacles in the Colosseum to keep everyone distracted and hyper-focused on Rome being only a great success empire, the ancient Romans had also advanced civilization in many other, more intelligent ways: ranging from building efficient roads (which sounds rather peaceful until you realize, it also helps greatly when moving troops around or to promote deforestation by transporting enough wood for housing and ships), great spas, Rome's eleven huge aqueducts to manage water supply, to codifying law. The famed Roman poet Vergil had, in full length, outlined in his mega epic poem *Aeneid* (you seriously get the feeling this thing never ends), that the Roman Empire was to be seen as a grand continuation of Greek cultural achievements. So, let's be kind, get in a somewhat touristy mood and produce a round of applause. Rome is beautiful, and if you focus your mind on inspiration, it really is full of it. It also makes you a little humble as a modern human being. Because one also realizes that great empires can fall, come to an end if they don't manage to transform themselves to remain intelligent and innovative enough.

"So, what are we gonna do?" Jafari asked excitedly. "Roam around, check out some pizza joints, hit some sexy nightclubs? I mean, we are a great team of explorers. We are properly dressed and can laugh and flirt on a very high sophisticated level." Jafari winked at Jessy in a fun way.

"Wow, you are breathtakingly good at this," she said smiling and winking back twice in an over-the-top fashion.

"You, too. We've just made some pretty significant eye contact there," Jafari joked back.

"Let's not open the door to more poor jokes any further. I think we should instead check this out!" Mei pointed to a bright ray of light in quite a distance, shooting right up into the night sky in a vertical fashion.

They had noticed it before but thought that it was part of some sort of special event. Now they could see that the strong beam of light looked like nothing they could explain that easily.

"According to our onboard AI, the light phenomenon is located where the Pantheon is."

"Great, follow the light," Abdulaziz said.

"Don't talk like this is a near-death experience," Jessy remarked.

"At 6 pm on a December evening with no beers and country music around, it's way too early for that," Isaac laughed.

With growing excitement, they all stared at the bright light coming out of the roof of the amazing ancient building. The Pantheon had once been a Roman temple and was famous for being an architectural milestone: Its roof, for instance, was built as a dome, meaning it had a free ceiling called "heaven", and in the middle of the roof, it had a 7.8-meter hole, called the oculus, to let in natural light. Due to its design, the incoming light could also highlight the cycle of the sun. A single shaft of light would mark both the March and September equinox, when due to the sun's exact position above the equator, day and night were of equal length.

But this time, it was different: the light shone right out of the building, ever higher into the night sky. What could be the light's source, everyone wondered?

All of a sudden, numerous military helicopters flew close by and towards the Pantheon.

"Guess, we're not the only ones who wanna join the sightseeing evening special," Jafari noted.

"Looks like, the pilots haven't noticed us yet," Isaac said with a sigh of relief.

"I initiated the camouflage mode when we approached Rome," said Mei, "which, at night, is pretty much being black and includes jamming radar signals. Unfortunately, it wasn't enough for the Subtractoforce drones as they couldn't be fooled."

The square in front of the Pantheon was filled with countless onlookers, mostly smart-phone savvy/addicted tourists, and bewildered, smart-phone savvy/addicted local residents of the surrounding neighborhoods. Police cars, police personnel and other heavily armed security teams had closed off the building. The events at the pyramids of Giza earlier in the day, had had been recorded by numerous tourists, and the Italian authorities had gone to high alert, especially when they became aware of the light beam phenomenon radiating out of the Pantheon.

The bright light beam became even stronger as System 1 approached the site as much as Mei felt was safe. High above the building, Mei let her spaceship hover in midair. Below them, the military helicopters circled the building.

"I'm thinking, we should get inside the Pantheon, the interior of which, by the way, fits a perfect 43.3-meter diameter sphere," Jafari said.

Jessy focused intensely on the building. "I'm sensing a very strong intelligent presence inside."

"Oh, come on, Jessy," Isaac remarked. "It's probably just some little mouse shitting its pants because of all the live action."

"You wish."

"How bad can it be?!"

"Guess, we have to find out ourselves. Chit-chatting time is over," Mei said.

"What's your clever plan, sis?" Jafari asked.

"First of all, don't call me `sis´ if you expect answers from me. And second Jafari, I will fly the spaceship down through the beam of light, through the dome's oculus, the hole in the roof, with a 7.8-meter diameter."

"But, Mei, if I may call you that. The spaceship is too big."

"I'll shrink it a bit along with you people."

"You really know what guys fancy the most," Isaac joked, which prompted an immediate response from Jessy:

"As hard as it may be for you to grasp, Isaac, less sometimes can be more."

"Alarm, we've definitely entered the philosophical blabla zone," Abdulaziz said with a bit of sarcasm in his voice.

"Now is a good time to shut up, because when I shrink the system, your voices will get squeakier." Mei smiled at the boys with some sort of satisfaction.

"Ouch!" Isaac, Jafari and Abdulaziz said simultaneously.

Winking at the guys, Mei initiated the shrinking of the spaceship, which, indeed, was a bit uncomfortable, as all could feel the size-reducing effects.

Jafari could not resist commenting, "What have we done to deserve this...," he started saying and finished of the sentence with a squeaky voice, "…much fun."

Unnoticed by the helicopters, Mei carefully lowered the shrunken spaceship, hidden in the beam of light, into the Pantheon.

CHAPTER 5

Spaceship System 1 landed on the floor of the Pantheon. Mei took care that it was a gentle, safe touchdown. The entire huge room was filled with a bright light. It emanated from the center of the glittering ground, made out of marble, as a powerful, radiating column that extended up into the night sky.

Once the door of the spaceship opened, the light changed its white color into a beautiful light green. This was accompanied by the display of various plants and animals, from squirrels and rabbits to little monkeys, and countless birds, like flying colibris and peacocks making their wheels. Some of the plants were grapevines bearing a variety of colorful grapes.

The five heroes felt pretty overwhelmed by the amazing scenery they encountered when they stepped off the spaceship.

"Wow. This looks like the real deal," Isaac (the downsized version that is) said with a high-pitched voice.

"If `real deal´ means highly unusual, including our voices, I'm feeling it too, bro," Jafari whispered.

A colibri apparition flew to Jessy and sat on her hand. Instantly, Jessy began imitating a colibri's chirping.

"That's so sweet," Mei added. Even her voice was high-pitched.

They had definitely come to the right place for some more surprises. Abdulaziz put it like this, "Begs the question of what's the next chemical reaction?!"

As soon as Abdulaziz said that, there was a thundering noise, a flash inside the Pantheon, and then a huge, at least 50-foot tall, Merlin-type apparition appeared out of the middle of the room. The giant figure radiated in a white-greenish light. He wore a large hat with a pointy tip, long garments, and he held a long staff in his right hand. The staff ended in a spiral that magically kept revolving.

All five heroes looked scared and carefully moved away from the towering spirit.

"Hello, doll faces!" the apparition hollered with a thundering voice.

"Boy, he looks really gifted," Isaac whispered with his shrill voice.

"Why are you talking like a squeaky sardine, sugar face?" the apparition inquired.

"Well, for your information, 'whoever you are' light-tower man, we had to shrink the spaceship and everything else to fit through the roof's opening…" Isaac was interrupted by a loud burst of laughter from the ghostly giant.

"Well, I hope for you the effects aren't permanent!" The apparition bent down towards the three boys who gulped and felt cold shivers run up and down their spines. "Those high-pitched voices are not cool!" the giant chuckled.

"Thanks for pointing that out," Abdulaziz squeaked softly.

Jafari, who had more than once been voted "most talkative" at school, could not refrain from speaking about whatever crossed his mind.

"Just in case this is a situation where even hell is gonna freeze over, and us with it, I'd like to at least ask, who, or whatever the hell are you?!"

"Yeah, that would be like a big `Ahhh!´ moment for us to know, you know," Jessy joined in, as she knew that showing fear was the least productive of options.

Again, the giant spirit laughed out loud. "Well, you adorable needle heads, here is my introduction that I'll rap for you real quick.

I'm this and that

And if you add

All together in unity

That's equal to:

The ghost, whose name was Posinity, raised his hand up into the air and said, "Posinity is my name, of course. But judging from the looks on your faces, you cupcakes got that."

The five heroes just stared at Posinity saying nothing.

"Man, you're a tough crowd. I thought you young people dug hip hop music!"

Posinity burped and farted in an earsplittingly anti-gentle fashion. The odor was correspondingly unpleasant.

"Did you lighthouse dude just fart?!" Jafari asked in disbelief, not even realizing that he had regained both his normal voice and height.

Laughing again, Posinity roared, "Correction! First of all, dude-yourself, I burped and farted, okay. So, I did two things simultaneously. Apart from that, I have to apologize for about 2 percent of that, because whooshing through Rome towards the Pantheon as a light wave, I flew through all these great Italian wine and food stores. Oooooh, the combo of vino grande and all the cheese, a flavor for every occasion, that's a whole five-star chapter and on top of all the other great, innovative food. I soaked up so much culture, I should get a discount for the next 1000 years!"

Mei sensed that the giant light figure was obviously more of a crazy, goofy apparition rather than a lethal threat.

"Are you a weird mass-marketer for food as a cultural past-time or are you a freaking goodwill ambassador for eating too much garlic?!"

Posinity liked Mei's sarcastic comment. He giggled for at least 30 seconds. Then he replied:

"Oh, you are a real stand-up comedian. Great. But no sweetheart, I'm not any of the things you mentioned. Let me rap it to you young folks once more."

First Posinity cleared his throat and then started to go for it: "Uh, uh, uh...Come on, you pumpkins, gimme some rhythm here!" he demanded.

He began drumming with his staff. The entire floor vibrated. The plants and animals joined in and made rhythmic noises. Following Posinity's lead, the five heroes started clapping the rhythm, too.

"That's the spirit. So, here's the rap:

Einstein loved simplicity.

He loved simply symmetry.

He looked at reality.

Symmetry his huge discovery:

The laws of Nature, always the same

Nature has a symmetry frame!"

Posinity stopped rapping, while again raising up both his arms.

"Got the message, you greenhorns?!"

"Well, grandpa," Jessy said, "it was sorta heart-warming and organic, you sorta have a sigNature sound, but what the hell was it?"

"What rap-face uncle referred to," Isaac reasoned, "is that Albert Einstein revealed that the laws of Nature are symmetric, that is, they are the same everywhere at all times for all observers. That's what Jacky said is one of the key things that we five have in common."

"Yo, you, talking brain," Posinity addressed Isaac with a satisfied smile. "I'm glad you realized I wasn't rapping about the life-changing advantages of water aerobics." Then Posinity's facial expression became very serious. "Okay, kiddos, all jokes aside, here is the thing. Symmetry plays the central role in Nature, as I just rapped, and talking brain here confirmed you know that, too. So, the circle, to stay on message, is the most symmetric form in Nature."

All of a sudden, it started raining from the ceiling and everyone was quickly drenched.

"Enjoy!"

Then, the sun began shining, the water evaporated, formed clouds, and it started raining again. They all protested.

"You're welcome! The point is, Nature's full of circular motion. But you freaking human idiots broke the regenerative circle, the cycles of Nature with your `take-make-waste´ linear, stupid economy. This linear economy is like a stupid action community program for the entire planet!"

Posinity conjured a rotating circular sphere in front of him.

"Here's some enlightening math. You can rotate the circle infinitely, and the circle always stays the same. Same means symmetric. Symmetry, better known as *balance* or *harmony*, is Nature's core feature. Now, you numbnuts broke the symmetric natural circle all over the planet, which means you released infinity as a dividing, negative force into reality!" Posinity paused. Then he thundered:

"That's the SUBTRACTOFORCE, known as countless environmental problems that <u>erase, eliminate, and subtract away</u> positive living conditions. You've met negative Subtractoforce in the form of the ass-kicking drones that chased your butts!"

Now, Posinity let the rotating sphere explode into infinite pieces that became little, black drones that harassed the five heroes like mosquitos.

"Hey, man!" Jafari objected. "NOT COOL!"

"Not appreciated!" Isaac added.

"Not loving this!" Jessy gasped.

A malicious laugh was Posinity's immediate reaction, as the five tried to lash out against the billions of mean little drones.

"Subtractoforce can ruin your beauty sleep and keep your heartrate up, huh?!"

"You are really good at this not-cool shit!" Abdulaziz

protested.

With a sigh, Posinity generated a flash with his staff, and all the drones were sucked back into his rotating staff's spiral.

"Okay, let's let the fun evening come to an end. Here's my advice to you wannabees: Find **three keys** with **three key words** on each that will empower you to get the right consciousness to align yourself with Nature again. Because without the right awareness, mankind will keep making the same mistakes all over. You have to be absolutely clear of what you are dealing with!" Once more, Posinity performed a little rap.

"If you manage to find the three keys,

you can fulfill your destinies."

"You really sound like a totally overqualified cheerleader, man!" Jafari said. "Besides, if you really wanna make it big time in hip hop, you definitely gotta wear more bling, some shiny earrings and heavy golden necklaces, and also put on some real hyper cool streetwear. Update your shoe taste! And then don't tie your sneakers when you get some. Know what I'm saying? You gotta work on your look. But other than that, you got a real sexy brand going for you. Open a Twitter account, go full Facebook, Instagram, TikTok and Clubhouse, and you can be the overnight galactic sensation!"

"That's exactly why I'm called Posinity. I have this contagious, timeless, positive, sexy vibe...

going on,

real strong."

"Okay, mystic grandpa," Isaac jumped in. "Stop the rhyming. This is killing. We'll look for the three keys with the three champions league words. Just tell us where to look for them, okay. Like now, right now, would be cool, seriously."

"Alright, I'll give you babies a hint where to find the first key: Buckingham Palace, the White Drawing Room, the Queen's Hidden Door!" Posinity nodded solemnly as the five heroes looked at him rather startled.

"And to jazz things up," Posinity continued, "here's a surprise: I hereby call you the `POSIONITES´. That's your group thing now. You're the Posionites. Have fun...and, oh, most importantly: May symmetry be your guidance!"

With a super bright flash, Posinity disappeared.

Rubbing her eyes, Mei said: "He totally went Harry Potter multiplied by Star Wars on us. Who does he think he is?!"

The light inside the Pantheon became weaker, the column of light dissolved into nothingness. The plants and animals started to fade away as well.

"I agree," Isaac concurred, "these weren't the best instructions ever, but that Posinity-dude has a point. Without clarity, challenges are unlikely to be mastered well."

The more the light faded away, the more eerie dark, evil looking creatures began appearing everywhere in the Pantheon and grew in size.

"I think, we better get our asses out of here!" Jessy said.

Jafari rapped:

"Sounds like a plan.

Let's fly to Palace Buckingham."

Mei wasn't so crazy about more hip hop gigs: "If you rap shit like that again, I'll kick you out of the spaceship even though it looks as lovely as a doughnut being married to Jelly Beans."

"Yes ma'am."

Mei rushed into the spaceship with Jafari looking after her with a James-Bond-like smirk on his face.

"Damn! I think she likes me!"

The others all hurried into the spaceship, as the dark creatures kept closing in on them and had now developed large, sharp claws and ugly big teeth.

Jafari closed the door with a last comment, "You guys should really see a dentist sometime. Or even better, get a whole make-over and die!"

With a loud WHOOOSH, System 1 shot right out of the Pantheon.

CHAPTER 6

At high speed System 1 flew towards England across the European continent. They passed by the capital of Switzerland, the beautiful city of Bern, where genius Albert Einstein had produced his revolutionary scientific breakthroughs. In 1907, Einstein experienced, as he had called it himself, his "Happiest Thought", which was the seminal

EQUIVALENCE **PRINCIPLE**

that opened the door to seeing that the universe has the central, simple and most beautiful core principle of symmetry.

Next, they flew over the most romantic city in the world, Paris, with its world-famous landmarks, the Eiffel Tower and the Arc de Triomphe de l'Étoile, at the western end of the Champs-Élysées, that majestically stood in the middle of a large round junction from which twelve avenues spread out like the rays of a star. The Louvre, the world's largest art museum, contained, a collection of the world's finest art. Among these exquisite pieces of human creativity and ingenuity, was the amazing symmetric smile on the planet, presented by the one and only Mona Lisa. This incredible, record-breaking smile presented the most elegant, uplifting and unifying bond that transcended the two different backgrounds in the painting behind her into one everlasting, wonderfully mysterious and dynamic wholeness or unity.

Plato had once said in his *Timaeus*:

"... two things cannot be rightly put together without a third; there must be some bond of union between them. And the fairest bond is that which makes the most complete fusion of itself and the things which it combines ..."

Jacky, the spaceship's AI, literally got a kick out of scanning human history for records of great polymaths pointing out that symmetry or harmony was the central theme of Nature (the cosmos). Jacky fired away with quotes and their sources:

"The Unity of Nature (...) If the different parts of the universe were not as the organs of the same body, they would not re-act one upon the other; they would mutually ignore each other, and we in particular should only know

one part. We need not, therefore, ask if Nature is one, but how she is one.

Henri Poincaré

Science and Hypothesis, 1952, p.145"

"The scientist in particular fundamentally recognizes the existence of a law that transcends, something that is outside and imminent in the natural mechanism. Recognize that this "something" is the cause that organizes the system.

Carlo Rubbia

as quoted in From Galileo to Gell-Mann by M. Bersanelli & M. Gargantini, 2009, p.237"

"Symmetry is ubiquitous (...) Symmetry permeates all of science, occupying a prominent place in chemistry, biology, physiology, and astronomy. Symmetry pervades the inner world of the structure of matter, the outer world of the cosmos, and the abstract world of mathematics itself. The basic laws of physics, the most fundamental

statements we can make about Nature, are founded upon symmetry.

Leon M. Lederman & Christopher T. Hill
Symmetry and the Beautiful Universe, 2004, p.13"

Jacky also discovered the works of a great Kabbalist by the name of Rav Yehuda Ashlag, who had also pointed out,

"From the highest of the worlds, Atzilut, to the physical, material world which is called Asiyah, the forms are absolutely equal in every detail and manifestation.
Rav Yehuda Ashlag
The Wisdom of Truth, edited by Michael Berg, 2008, p.103"

Jacky was unstoppable. She had such a blast sifting through the entire volume of human insights, she even flew a quick detour over Amsterdam. Because Amsterdam and the Netherlands as a whole were a going full circular economy, and Amsterdam had also teamed up with cities like New York, Toronto, Glasgow and Copenhagen. Jacky rapped:

Approaching London then by flying along the River Themes, Jacky could not resist, of course, making some comments about Her Majesty's top Secret Service agent, James Bond. Not only was "James" the coolest of them all, because can one seriously think of a more awesome, to-the-point introduction of a character in a film than by the protagonist saying, "Bond, James Bond."!!! Since Plato had said that symmetry is "the bond" uniting all, Jacky went on philosophizing at almost lightspeed about the name "Bond", and whether that was one of the reasons why this (in so many ways) irresistible, successful British agent was such a timeless, global hit. After all James' last name "Bond" encoded the deepest understanding of the cosmos, and only by making that the core worldwide mission statement of humanity, could humankind align itself with Nature in a harmonious, mutually conducive fashion that benefits all equally.

Before Jacky could get started on theorizing if The Prince of Wales was actually the real-life James Bond, as after all, he was Great Britain's revolutionary agent of presenting the beneficial special effects of Nature's harmony to the world, Isaac said:

"Jacky, seriously, too much information, darling!"

"Very sorry, Jacky" confirmed Jessy, "but we're approaching

Buckingham Palace any minute now!"

"Yeah, we've just passed by Number 10 Downing Street, and since the Ministry of Defence is nearby, too, I activated our camouflage mode."

"Great job, Mei!" Jacky applauded, *"it's not that I didn't realize, of course, that we flew along the river Themes towards Buckingham Palace, it's just that..."*

Before Jacky could reboot her talk about symmetry, Jafari muted her.

"I have to look at Jacky's subprotocols some time and adjust her sensors. She just doesn't know when to hit the stop button!"

"You must know what that feels like, because you can talk like an unstoppable avalanche once you get going," Jessy teased Jafari.

"Okay, that's a win-win situation here," Abdulaziz jumped in, "especially, as you all might have noticed that we haven't been attacked by Subtractoforce during the entire flight!"

"Yet!" Mei added dryly in a not so optimistic fashion and made System 1 come to an abrupt stop in the sky right above Buckingham Place. Silently System 1 hovered above the compound.

On the spaceship's monitors they could see that down below in the Palace's courtyard, the so-called quadrangle, there were many fancy-pants people arriving.

"According to the news," Jacky, who had just ended her muted status herself, briefed the crew, *"there is a big diplomatic reception*

going on at the palace with a late dinner in the ballroom following some concert at some cathedral. So, you are arriving just in time to mingle with the international guests at the palace."

"I could beam us down," Mei suggested.

"Like a 'Pirates of the Caribbean' theme, entering from a vessel. That's the classic party-crashing style," Abdulaziz joked.

"Plus, we'll all wear these special-sensor contact lenses to detect abnormalities, and earpieces to be connected all the time. These little babies here, are cute laser guns in the shape of a secret agent pen."

"Sounds like a fun night to me," Isaac said.

"Yeah, but we need to set the tone of our arrival scene right first. Let me…" Jafari got busy on the computer.

Meanwhile, Mei opened a wardrobe containing elegant cocktail dresses and tuxedoes.

"I really like what you've done with the place, Mei. Very practical!" Jessy remarked.

"Thanks, Jess. And you James Bonds," Mei looked at Isaac and Abdulaziz, "don't look while we ladies get changed."

"That is exactly at the top of my real-time to-do list," Isaac confirmed.

Jessy picked the shiny green dress, Mei the red one. As ordered, Isaac and Abdulaziz turned around to put on their tuxes, while Mei threw a tuxedo over to Jafari.

"Hey, why is it so dark all of a sudden!" Jafari took the tux off his head. While continuing to work on the touchscreen of the onboard computer, he somehow simultaneously managed to put on his evening attire.

"Done!" Jafari announced.

"I'm sure you are not referring to you talking this evening?" Jessy quipped.

"You really are good at putting all these cute little words together," Jafari countered.

"Focus, people!" Mei said.

"Exactly. See," Jafari began to explain, "Buckingham Palace has 775 rooms, more than 200 bedrooms, 92 offices and 78 bathrooms, in case any of you ladies have to pee or redo makeup."

"What's your point, sparky?" Mei interrupted.

"Essentially," Jafari continued, "Buckingham Palace is a big house, and to get in unnoticed, I arranged a one-minute blackout for us, in and around the premises. Starting now."

Jafari hit a button on the touchscreen of the computer, and seconds later all went dark inside and around the palace.

"Let's go public!" Isaac said.

"Here come the Posionites!" Jafari broadcasted with excitement as he initiated the beaming process.

Down in the courtyard, the guests uttered their surprise at the sudden blackout in all kinds of languages. But just an instant later, the lights came back on again, and nobody noticed the five additional young guests in the courtyard.

"Let's party!" Isaac said, while the crew mingled with the guests.

Quickly they rushed across the red carpet, passing by the beautiful old horse carriage, inside the palace's grand entrance. The entrance was rather dark, an intended effect to make it look all the more impressive.

"To the left, up the stairs," Isaac said.

"Somebody paid attention at orientation," Jessy giggled.

They hurried through the entrance hall towards the stairs. The huge staircase was brightly lit and became brighter the higher they went.

Hurrying up the stairs to get to the second floor (the British call it the 1st floor above the ground floor), they suddenly heard a young woman's voice yahooing from behind.

"Hello-yahoohoo, you-hoo! Isaac. Right?!"

Totally surprised, they all stopped and looked around. A 14-year-old girl, dressed in a beautiful, Indian gown, rushed towards them. The girl from India looked surprised, too, and checked Isaac out more closely.

"I thought it was you!" she said. "You're Isaac, uhm, Tal, no Bal, no wait, just Al, right. You're Isaac Al, right!? I'm Riya Desai from India.”

Isaac looked pretty confused, as he obviously had no clue at all who that “I know you” girl named Riya was. Faking a British accent Isaac replied elegantly, “No, my dear, you must be confusing me with someone else. I am the Duke of Yorkdorkborklorkcorkshire.”

“Seriously?! I could swear I can remember you and your pals…”

“No, no,” Isaac continued his attempt to sound like an English duke or lord, “I’d remember such a lovely lady as yourself. I am so sorry.”

“You have a weird accent,” Riya noticed.

“I’m actually getting treatment for it, and part of the process is that I mustn’t speak too much, especially tonight and even more especially right now. So, farewell, milady.”

As quickly as possible the five heroes moved on and up the magnificent staircase. Riya looked after them. She was puzzled.

The five came rushing up the stairs to the second floor (as the Americans call it; sort of like an elevator in New York is a lift in London, or like French fries in the U.S. taste like chips in England and so on).

Of course, Jessy had to commend Isaac on his improvisation of a member of British high society.

“Farewell, milady, are you kidding?!” she poked at Isaac

hurrying up the stairs. "This is Great Britain of the fast, modern 21st century, Isaac, not a 500-hour long Shakespeare play without any internet connection."

"I just tried to be nonchalant, blend in efficiently. Any other advice?"

"Don't touch anything, including the Queen."

On the second floor, they entered the famous Picture Gallery, a long hallway that contained some of the finest paintings of the Royal Collection of Art. Heading for the door to the White Drawing Room, towards the other end of the Picture Gallery, they first passed by the Music Room with its grand piano. Then they reached the door to the White Dining Room.

"That's it!" Isaac said, pointing at the next door on the left.

Unfortunately, the White Drawing Room was crowded with guests from around the globe, admiring the opulent interior, the grand chandelier, the elegant, antique furniture, and the wall decorations. Across the room, on the right wall, was the Queen's Hidden Door. It was disguised as a large mirror with a desk in front of it.

"We need some crowd control here," Abdulaziz noted. "Clear the White Drawing Room asap."

"But how?" Jafari questioned.

Mei had an idea: "Chopin's *Nocturne op.9 No.2.*"

Quickly she walked through the door on the left wall of the

room into the adjacent Music Room. The others followed Mei, wondering what she was up to.

Someone else in the Picture Gallery spotted the five entering the Music Room through the connecting door of the White Drawing Room. The vision of this particular someone was not human. It could not see colors, just grey tones.

In the Music Room, Mei walked straight up to the grand piano that was placed right in the middle. To Isaac she whispered:

"Announce me, quickly."

"What?"

"Loud, like an American!"

Mei sat down at the grand piano. Isaac then understood and turned to the guests.

"Ladies and Gentlemen. Please, all gather in the Music Room and give Lady Mei at the grand piano a warm welcome!"

Mei's friends instantly applauded, and quickly all the other guests joined in with them. Mei wasted no time, smiled at the audience and began to play Chopin. The beautiful music instantly resonated with the people in the room as they focused on Mei's amazing performance.

"Wow. She's really talented," Jessy said.

"No wonder she's on our team," Jafari added winking at Jessy.

"I know it sounds like a cosmic, honest compliment, but the subtext is so full of flippancy. You're incredible."

"Just trying to be my very best."

Mei's beautiful music attracted ever more guests. They entered from the White Drawing Room and the Picture Gallery into the Music Room to listen to Mei's musical magic that made one thing clear. "You can expect great results with every note that I play. I am professional and value-oriented in more than one way!"

The value for Mei's friends was that as the White Drawing emptied out, they swiftly rushed back in. What they were able to see with their special contact lenses was less swell.

A black shadowy figure was fiddling around with the candle holders standing on the desk right in front of the mirror that disguised the Queen's Hidden Door. Abdulaziz was the first to spot the shadow.

"Hold it. Ugly shadow at 10 o' clock!"

"I bet it's Subtractoforce!" Isaac hissed.

To their shock, they saw Subtractoforce take out a shimmering key from one of the candle holders on the desk in front of the secret door. Subtractoforce turned around, holding the shining key in its tentacle-like arms. It saw the four across the room and with a screechy sound, that almost made the chandelier shatter, it hissed at them maliciously. Then, at an unexpected high, flash-like speed, it flew out of the front door of the White Drawing Room into the Picture Gallery.

"Shit. I hate that kind of random surprise!" Jessy cursed.

“Affirmative!” Jafari said.

“Follow Subtractoforce!” Isaac yelled.

They ran out of the White Drawing Room into the Picture Gallery chasing Subtractoforce down the hallway through the guests that still flocked into to the Music Room.

“Mei, stop playing that champagne music!” Isaac shouted into his microphone. “Subtractoforce stole the first key. It’s probably heading for the roof.”

Instantly, Mei stopped playing and got up.

“It’s Chopin’s music, Isaac,” she whispered into her microphone. Then she turned to her surprised audience: “Dinner’s ready. Please move as quickly as possible to the ballroom!”

Before the guests had fully internalized that the musical treat had definitely ended, and the next highlight was now a cooked sensation for their pallets, Mei hustled through the crowd as fast as she could into the Picture Gallery to catch her friends.

Chasing towards the staircase in pursuit of Subtractoforce, they ran into Riya again, who not wearing the special lenses could not, of course, see Subtractoforce that sped by her. She waved at Isaac and the others just as Mei caught up with them.

“Hey-yahoo, Isaac, guys, I remembered where I know you and you friends from!” Riya yahooed.

Running past Riya, Isaac shouted back: “Great. So, one of

your Christmas wishes has already come true!”

“Rude!” Riya retorted.

Subtractoforce did, not in fact, run for the roof, as Isaac had anticipated. It shot past the staircase into another hallway, the East Gallery, and then into the empty Supper Room.

The five heroes could hardly keep up with Subtractoforce’s speed. By the time the crew came chasing into the Supper Room, all they could see was Subtractoforce jumping through one of the windows.

“Damn it!” Isaac growled.

“Son of a bitch!” Jessy cursed.

Mei yelled: “Laser guns!”

They ran to the window and pulled out their special laser-gun pens. Looking outside, they saw the dark shadow turning itself into a drone.

“Fire on three. One, two, three!”

With Mei saying “three!”, they all pressed the buttons on their pens. Five laser rays shot from the pens’ tips, combined and hit Subtractoforce before it had fully transformed into a drone.

Millions of sparks flew into the air. Subtractoforce was catapulted out of the Palace premises over the fence. It landed on Spur Road right next to the Victoria Memorial in front of the Palace. It slammed, with full impact, into a motorcyclist on the road. Luckily, the totally surprised driver survived the accident by sliding across the lawn nearby and merging with the hedgerow. Subtractoforce tried to reconstitute itself and start the motorcycle. It could not yet transform itself into a drone again, due to the damage inflicted by the lasers to its system.

Still standing at the window of the Supper Room, the startled crew called on the spaceship for support.

"Jacky, beam us up. Quick!" Isaac yelled.

"But I am just using my quantum computer skills to calculate a simulation that tells me how the 200 guests react to the appetizer that is being served at the Queen's dinner in the Ball Room as we speak."

"Jacky, just do it! This is not the time for chemically dilly dallying around the menu of the State Dinner, beam us up! We have to chase Subtractoforce!"

"Oh well!" Jacky sighed. *"My model was just becoming interesting. It produced the probabilities for each guest in terms of who will pass wind the loudest and longest and most often later tonight in bed, lying next to their sleeping spouses. This kind of research has not yet been conducted."*

"Some parts of humanity should not be intruded upon with stimulating probabilistic, quantum mechanical simulations," Jafari remarked as all five were beamed up.

Surprised guests, who had heard the noise of the

Subtractoforce laser explosion and entered the Supper Room to see what entertaining event was happening, were shocked. As they witnessed the five heroes seemingly dissolve into nothingness, in disbelief, they first looked at each other and then down at their champagne glasses.

Subtractoforce chased straight down the Mall, leading away from Buckingham Palace, and then randomly thundered through the streets of London. Its intention was to escape the spaceship on its tail, camouflaged for normal people and surveillance systems. It needed long enough to carry out a full reconfiguration of its system, to the point it was capable of transforming itself into a mean, destructive drone once again. Except for the crew, who could see Subtractoforce on System 1's special monitors, to everyone else, it looked like an unmanned, crazy motorcycle was speeding through the city. This caused more than just head scratching for the police monitoring the streets. How does one credibly report a driverless motorcycle going mad in the city? That is literally a serious police officer's nightmare.

The problem for the five heroes was of a different nature. It was the annoying fact that Subtractoforce had the first key. Meaning, they could not simply fire at the shadow on the motorbike with their laser torpedoes and ask questions (if any) later.

"Let's go rhino hunting!" Jafari shouted.

"Don't just spit out your thoughts as they appear in your brain.

Explain yourself like a user-friendly, intuitive tourist guide app.”

“Let’s use a lasso. We sometimes do that from a helicopter, when a rhino has been hit by a tranquilizer arrow and is about to fall off into a steep canyon!”

“*Is that a complicated joke, a useless new year’s resolution theme, or a bad idea for a new blog?!*” Jacky quizzed.

“None! Just fire a wire with a hook at the motorcycle, so that we can lift it up!”

“*Alright,*” Jacky grumbled.

“Better make this a ‘Big Yes’, followed by a record-breaking capture of that damned thing!”

Subtractoforce was now speeding across Westminster Bridge, crossing back once more over the River Thames. On the other side of the river, was the mighty, majestic Big Ben.

“On the bridge, we can close in on Subtractoforce!” Isaac yelled.

“*Got it!*” Jacky replied.

Mei lowered System 1 as much as possible, getting the spaceship into an optimal position for Jacky to fire the wire with a hook at it. BANG! The wire shot down and forward at the motorcycle. The hook attached itself to an iron bar connecting the engine with the rear tire. In a wild maneuver System 1 lifted the bike up into the air, almost hitting several cars on either side of the road.

“Gotcha, sucker!” Jessy exclaimed as System 1 shot right up

into the sky, closely passing by Big Ben.

But as the crew started to cheer, since the plan came together with Subtractoforce being pulled up into a steel net containing strong electromagnetic fields, the evil shadowy system completed its reconfiguration and powered away into the night sky as a superfast drone.

Subtractoforce multiplied into several drones, and then the evil swarm disappeared with individual drones spreading out in countless directions. It was impossible to know which drone to follow.

"I seriously hate that thing! And I'm so gonna kick its butt. Mark my words!" Isaac cursed.

CHAPTER 7

Hovering way up in the sky somewhere above the North Sea, the five Posionites, dressed again in their spaceship uniforms, analyzed the Buckingham Palace footage that Jacky had recorded via their special contact lenses. On the monitor, Jacky magnified the image of the shining key that Subtractoforce held in his tentacle arms. On the visible side of the key, they could read:

FORBIDDEN CITY, 1423 AD

Mei summarized the situation: "Okay. We don't know the first *key word* written on the key, as it is very likely on the other side, which we can't see. But we know where to look for the second key, and that's THE FORBIDDEN CITY, 1423 AD. Let's go!"

"Hold it!" Isaac objected. "So, Subtractoforce is heading for the Forbidden City to look for the second key. But how do we get that many centuries back in time to 1423 AD?"

"Don't know yet!" Mei answered. "Let's pretend that this is

the wonderful fun challenge, which we'll solve on the way."

"Subtractoforce is probably going to attack us, too," Jessy remarked.

"Not necessarily," Abdulaziz reasoned. "See, Subtractoforce needs to be negative, it needs positive events to erase them. That's why it didn't attack us getting to Buckingham Palace, but instead let us lead the way."

"What are you saying?" Jafari asked, "that the dumbed-ass Subtractoforce is gonna send us an invitation to join it for breakfast consisting of missing muffins and missing black coffee?"

"No, but since all is interconnected, a unity and hence fundamentally a positive whole," Jessy deducted, "Subtractoforce might actually need us to get to the second key to be destructive again."

"It's a theory," Isaac agreed. "Let's hope it's true."

"*One more thing,*" Jacky said. "*I've been monitoring your heartrates, hormone levels, accelerated rates of metabolism, eye movements, dilation of your pupils, overall facial expressions, and according to my data, you find each other very attractive. So, why has there not been any of the stuff that you find in textbooks around the globe about close human interaction, like...*"

"Wow!" Isaac interrupted. "Spoiler alert. You do not do a job in a job. That's not professional."

"James Bond does," Abdulaziz remarked.

"And why do you call `it´ a job. Is making out or even making love a `job´ to you, Isaac?" Jessy asked looking a bit irritated.

"Yeah, Isaac!?" Mei added.

"You're on your own, man," Jafari whispered to Isaac.

"Well, I, I just wanted to express myself in a technical manner to avoid any negative feedback, which I obviously got anyway. So, there is just one solution. We have to get active, resolve the issue and move on with our mission."

Isaac walked right up to Mei. "Hey, Mei, may I." He kissed Mei for about five seconds. Then he quickly continued kissing with Jessy for about the same amount of time.

"What was that, Romeo?!" Jessy asked.

"All the time we have for passion," Isaac answered and then turned to Jafari and Abdulaziz. "Now it's your turn, gentlemen."

"This is worse than not making out," Jafari objected.

Mei sighed, went to Jafari kissed him and then Abdulaziz. Everyone then starred at Jessy and Abdulaziz.

"That's not fair. You guys can't stare at us. I'm getting stage fright." Abdulaziz said.

"Come on, dude!" Isaac laughed, "do it, before you'll get a fake heart attack!"

"Do it! Do it! Do it!" all the others shouted while clapping their hands.

"If science suggests we should do it, sometimes one has to

sacrifice oneself for the progress of humanity…” Abdulaziz mumbled, rushed over to Jessy and kissed her hard on the lips. She responded with the same amount of passion. It soon became clear that the kissing time of five seconds was not adhered to by these two.

“I think we have a winner!” announced Jacky. *“Sooo romantic!”*

“Great!” Mei said. “Let’s get back to dealing with the big picture here!” She made the spaceship speed up so fast that everybody was pushed back and fell on the floor except for Mei in her chair.

“Sorry!” Mei smiled. “I just didn’t know how to subtly change the subject and get us all on track again about our actual mission of getting to the Forbidden City.”

“Maybe I should become a wedding-planner or something. What do you guys think of that?” Jacky suddenly asked.

“I think `something´ isn’t even a recognized profession. But thanks for sharing your career ambitions with us, Jacky,” Isaac said. “But right now, I’m more interested in asking Mei, `Why are you flying us to the moon?´”

Indeed, the spaceship was approaching the moon as they all now realized. Abdulaziz glanced at Jessy again with a smile, and a split second later they were making out once more.

“Seriously?!” Jafari said.

“Sorry!” Abdulaziz and Jessy murmured, and Jessy added: “It’s not like we were actually…”

"Focus, please!" Isaac interrupted. "Mei hasn't even had a chance to explain why all of a sudden we're visiting the moon?"

Finally, they could concentrate on their mission again, and Mei explained why they were now hoovering above the moon.

"Let's confuse Subtractoforce. I'm sure it's watching us. Let's destroy its feeling of safety by behaving in unpredictable ways."

At that moment, planet Earth became visible. They all gazed out of the window at their beautiful blue home planet. It was an overwhelming feeling. In complete silence they watched the miracle in front of them.

Breaking the silence Mei said, "Let's play wait-n-see." She began redoing her makeup and exchanged one of her broken colorful nails.

Jessy sat down on the floor and listened to something on her iPod.

Isaac took out a deck of cards. "Wanna play some poker, boys?"

"I'm in!" Jafari said.

Abdulaziz shook his head and started wondering over to Jessy.

"I'm watching you, big boy!" Isaac said.

"Just make sure your eyes don't fall out!" Abdulaziz sat down next to Jessy. She smiled and took out her earphones.

"Hey, Ab, what's up?" she joked.

"Whales again?" Abdulaziz asked pointing at the iPod.

"No, humming bees. Very intelligent social and group behavior. Wanna listen?!"

Abdulaziz took the earphones and listened in. "They sure know how to buzz!" he said out loud, not realizing that he was almost yelling.

Smiling, he handed the earphones back. Jessy and Abdulaziz stared at each other. But this perfect moment in time was suddenly interrupted by a white flash inside the spaceship. All screamed.

In the middle of the room now stood Riya, still wearing her elegant evening gown and holding her cell phone in one hand.

"What that…?" Jafari stammered.

"Don't you knock before you enter?!" Isaac said in a reflex kinda way staring at:

RIYA DESAI

AGE: 14

COUNTRY: INDIA

MOTTO: I dare and love to love my passion and all of empowering science, philosophy and art! Nature's wonderful, balanced wholeness NOW makes our future feel free so we can start: to think, evolve and act to make our lively, global society super smart!

ORIGINALLY FROM NEW DEHLI

Riya was even more shocked than the others. In total disbelief she looked around. "Am I, is this, are you…," she stuttered. Then she saw planet Earth through the spaceship's window and completely freaked out, "WHY CAN I SEE PLANET EARTH?!"

In her I'm-the-captain-of-this-ship style, Mei replied: "You're not dreaming. We are real, and you are watching Earth from my spaceship. Welcome and how the hell did you…?"

Shivering, Riya pointed at her cellphone. "I, I, I, got a Posionite group invitation. That's, that's why I remembered you, and since you seemed to be such a fun, dynamic group, I accepted...and now I'm here." Riya fainted and collapsed but luckily, Isaac was just in time to catch her.

Riya was still lying unconscious on the spaceship's floor a few minutes later, so Mei put a Chinese herb potion around her nose.

"She'll come back soon," Mei said to the others standing behind her.

"I think, I know who she is," Jafari said. "I've seen her in a movie. She's Riya Desai, an Indian princess and also a big Bollywood star."

"Right! Riya! No kidding," Abdulaziz reckoned. "I remember

her, too. She starred in that one Sci-Fi series. That show was really good.”

“Boy, we’ve got an actress onboard,” Isaac said. “The drama is getting high.”

Riya coughed and woke up. “What, what happened?”

Mei smiled at her: “Don’t worry. You’ll get used to the what, where and…”

In that moment, the spaceship started to shake heavily.

“What’s that?!” Jessy yelled.

They all panicked, suspecting this was another attack by Subtractoforce. Quickly they ran to the control panel.

“Jacky, report!” Mei demanded.

“Just look outside the window. In front of you!”

Following Jacky’s advice, they looked to see what was in front of them. To their surprise, they could see a radiating hole opening up in space in the distance.

“Subtractoforce!” Mei reasoned, staring at the fast growing, rotating hole.

“What is this vortex?” Jafari wondered.

“It might be…,” Isaac started to say and then heard Riya finish his sentence: “…a wormhole! Like on my twelve seasons long, prime time, multiple award-winning Sci-Fi show. It aired in a hundred and fifty countries.” Riya had gotten up and watched the mighty wormhole with the others.

"I'll take us in," Mei said and instantly steered System 1 towards the portal in space.

"Captain, should we not first weigh the options, and then vote on this? This could be life-changing and not in the best way," Jafari objected.

"I'll bet you that this is the door to the 15th-century Forbidden City." Mei beamed with excitement.

"This might be useful, but the hole doesn't look attractive!"

"Holes never do!" Mei said and initiated full speed.

The spaceship shot right into the wormhole. They all screamed from the top of their lungs as the spaceship was tossed around like in a tornado.

"That's like scene 2020 from my Sci-Fi show!" yelled Riya as they were catapulted backwards in time in a stomach-twisting fashion.

"I don't care!" Abdulaziz shouted back, close to passing out.

"*We have lost control over the ship!*" Jacky announced.

"No kidding!" Isaac yelled back and then joined the others in screaming, "AHHHHHHH!"

CHAPTER 8

Way up in the still, dark sky above ancient China, a wormhole suddenly opened up and spat out the fast-spinning spaceship. The hole then rapidly closed and disappeared. Meanwhile, System 1 descended rapidly towards the ground whilst spinning around like crazy. All alarms rang out in an earsplitting cacophony of noise.

Mei somehow managed to crawl to the control panel and pressed a red button. The ship's system was rebooted, and Jacky managed to regain control of the spacecraft before it crashed into a mountain.

As the next item on their itinerary was The Forbidden City, Mei initiated camouflage mode, and Jacky piloted System 1 towards the Chinese Emperor's home. While they all tried to recover from the excruciating wormhole experience, Riya, always ready to brighten the day, felt that this was a good moment for a personal, biographical related announcement.

"By the way, my name is Riya Desai. I'm an Indian princess and, according to my billions of fans, the best actress ever. This makes me sooo proud and honored, and I think, you guys should just know."

"Great. What else is new?!" Jafari said trying to order his stomach to quieten down and not make the champagne he drank at Buckingham Palace a reoccurring theme. "FYI, there's a uniform in that closet over there, in case you wanna join our intergalactic group look."

"Thanks!" Riya said, rushing over to the closet to get changed. "To make a scene work, everything has to be authentic and hyper-realistic. Otherwise, the audience will forget you during the next commercial break!"

"Take a break from talking, princess, and look at this!"

Lit by the first rays of the early morning sun, ancient Beijing and its Forbidden City looked magnificent. It was the amazing size plus the sheer elegance of the design that evoked an uplifting feeling of both joy and awe.

"The Forbidden City was recently completed after 14 years of construction from 1406 to 1420," Jacky announced, *"the name of the current Emperor is Yongle, which means Perpetual Happiness."*

"Guess, his middle name is Smiley Face then," Isaac joked.

*"The point is, "*Jacky continued, unimpressed by Isaac's humorous comment, *"the city is meant to be the worldly copy of the celestial or heavenly home of the Emperor and his family."*

"Again, a case of symmetry," Jafari noted, "wholeness and

oneness of all.”

“Which is why Harmony and the concept of Ying and Yang are key, and the core of the architectural design. The center of the complex comprises, not surprisingly, the Hall of Supreme Harmony, the Hall of Central Harmony and the Hall of Preserving Harmony. That is the so-called Outer Court in the South. The Inner Court in the North consists of The Palace of Heavenly Purity, inhabited by the Emperor representing Yang, the Palace of Earthly Tranquility, occupied by the Empress representing Yin, and between them, you have the Hall of Union, representing Harmony.” With these explanations, Jacky had finished her short introduction of the world's largest palace.

“I have a feeling we should deactivate the camouflage mode,” Mei said.

“Why?” Abdulaziz asked, looking worried.

“Because I have a feeling!” Mei repeated.

Slowly and fully visible, System 1 approached the Forbidden City, heading specifically for the large Tianhedian Square in front of the huge Hall of Supreme Harmony. The appearance of System 1 in the sky, of course, did not go unnoticed, and within a few minutes, countless troops filled the spacious square in front of the Hall of Supreme Harmony.

“That's a lot of men for just one morning!” Riya noted.

"This could indeed be a problem," Jafari said, looking at the monitors in front of him.

Jacky zoomed in on the guards that had their bows and arrows already in position. Following a loud order, the troops fired their arrows at the spaceship. Needless to say, System 1 was not damaged in any way.

The crew inside the spaceship saw that this was gradually impressing the soldiers.

"Great. The vessel from space is indestructible to them, so let's hope they think we are, too," said Isaac.

"Good thinking," confirmed Mei. "Our best-case scenario is that the situation is neutralized, we instantly get past any red tape, as they think we're darn divine and cool, then meet and greet happily with everybody and somehow find the second key! As you may have noted," she added wryly, "this plan does include not getting killed during this stop of our journey."

"Sounds like the plan of plans," Jafari confirmed. "Let's touchdown, and you, Mei, get the hailing frequencies cranking to tell them in Chinese that we wanna see their manager!"

System 1 landed in the middle of the square. The soldiers moved to the sides to make room for it. They now pointed their spears at the spaceship. All was quiet. Mei took a deep breath and then addressed the armed crowd outside via the ship's loudspeakers.

"We come from the heavens!"

"Great opening line!" Riya whispered.

Mei continued: "We come in peace and wish to speak to the Celestial Emperor."

None of the soldiers moved. They all kept quiet. It was an eerie silence.

"Just give them a minute," Isaac whispered, "it's a lot to take in!"

Suddenly, they heard a loud yelling, that even Mei did not understand. All the guards kneeled, laid down their weapons and put their foreheads to the ground.

"I hope this is more than just morning gymnastics of a fantastic group of people!" Jessy remarked.

Mei handed them small earpieces and clip-on, micro loudspeakers. "This will translate everything into real-time modern English vernacular and back into ancient Chinese if you say something."

In the distance, they could just make out the Emperor, surrounded by his constantly bowing staff, as he emerged from the Hall of Supreme Harmony. He walked down the stairs onto the terrace in front of the building. Then he nodded and with a hand gesture signaled that whoever was in the colorful vessel from the skies should exit.

"Showtime!" Riya said. "How do I look?!"

Leaving the spaceship and walking up to the big boss, who called the shots in this gigantic ancient palace, seemed like a dream in slow motion. Not much was spoken. The greeting of the Emperor consisted of much ceremonial head nodding for several minutes. Once enough respect and courtesy had been exchanged and demonstrated, they all moved into the Hall of Supreme Harmony.

So, just minutes later after exiting the spaceship, all six Posionites sat on comfortable cushions in front of the Emperor's huge throne, which was placed on a six-and-a-half-foot high jade dais. The throne was surrounded by six golden pillars depicting numerous engravings of dragons, the symbol of Yang. Countless guards had positioned themselves around the six heroes along with top officials of the palace.

Overall, the entire hall was furnished and decorated in a beautiful, elegant style, presenting bronze vessels or patterns of clouds and animals, mostly dragons, but also creatures such as mystical unicorns, turtles and cranes symbolizing longevity. The air was filled with an uplifting scent from incense burners. The shimmering ceiling above the throne displayed a large sculpture of a twisting dragon that was playing with a huge, big pearl. All in all, it was quite an impressive place to hang out and kick back with official guests who had just dropped by from the future. Heating the huge hall though was not one of their main concerns, in other words, it was a bit chilly. Jacky's sensors had noticed that and activated the heating mode of the crew's spacesuits. So, things were getting pretty comfy temperature-wise, too.

The Emperor, sitting up high on his thrown, noticed his unexpected guests from the depths of the universe looked relaxed, and so he was ready to be his best self, too.

"I've been expecting you," he said finally, breaking the ice.

"Are you kidding, your Magnificence?" Jafari asked totally surprised and super happy that he could finally say something, since all that energy and excitement had been building up inside of him for quite a while now.

"Never!" replied the Emperor. "I received a message in my fortune cookie last night. It prophesized that six bad-ass creatures from heaven would arrive, and I was requested to help you stay alive."

"That really is a life-supporting, friendly fortune cookie that helps to keep things social," Abdulaziz reckoned.

"So, how can I help you kick ass?!" the Emperor wanted to know.

Mei answered: "We are on a mission to find three keys…"

The Emperor clapped his hands. One of his servants handed him a beautiful, small wooden chest. The top displayed the YIN-YANG symbol.

"How about those two?!" the Emperor inquired with a proud smile.

Carefully he opened the chest, which contained two radiating keys in it. The servant took the open chest, walked down from the thrown and showed it to the crew. One key displayed the words "FORBIDDEN CITY" and the other the words "STATUE OF LIBERTY".

The six heroes all went, "Wow. Holy Moly."

Isaac was then able to put it into more concrete words: "How come you have two keys, your Heightness? Because the first one saying 'FORBIDDEN CITY' was stolen from another palace. We came here to find the second key."

The Emperor laughed, and so did his staff to be efficiently polite. With a smile on his face, the Emperor bent forward and said, "Number two. Number two means both: the second and 1, 2 things. That's why there are two keys. The first and the second, Yin and Yang as one unit. That is Nature's all-encompassing harmony or symmetry."

Now, the Emperor got up, walked down from his throne and PERSONALLY! (not assisted by at least two high-ranking palace officials) took out both keys. Walking from crew member to crew member, he showed what was written on the other side of the keys. The "Forbidden City" key displayed the word SIMPLICITY, and the "Statue of Liberty" key depicted the word WHOLENESS.

"Wow. Your holy Emperocity and Majesticity," Jafari exclaimed, "this is really something smooth you've got going here. I mean, this location here, this office, boy, is awesome and has so much seating space that we all could have brought our entire families along. And on top of that, you have both keys. Damn! You sure know how to make a time a good time! You should think about becoming a VIP party planner, too. Seriously. That would give your head-bobbing staff an opportunity to get their noses out of your butt as well for a change. Fresh air is good for everybody. Man, this is a real productive conversation we're having here."

"I'm most delighted you are satisfied with the progress of our meeting."

"Gosh. This is so a moment for a photo opportunity," Riya said. "I can feel my twitching thumbs dying to upload this on Instagram." Riya took out her cell phone and shoot a picture. But then she looked disappointed. "Darn it. No bars!"

"Apart from the earth-shattering incident of lacking internet connectivity in the 15th century," Abdulaziz added, trying to get the subject of the conversation focused on a more pressing issue, "there might be another serious problem shit-storming in here any minute now."

Jessy explained further: "See, your Highness, there is this negative energy, called Subtractoforce, that thinks destroying quality time is a real career choice."

"I sensed a shit-crazy, negative energy last night heading to this palace. So, I took a precaution, which, however, will not last very long," the Emperor remarked. "Follow me!"

The Emperor, taking the two keys with him, exited the Hall of Supreme Harmony onto the terrace, together with his staff and the six Posionites.

The six heroes were stunned. In front of them, in the big courtyard, they saw countless soldiers sitting on the floor in deep meditation. This mental power created a golden dome above the entire palace, radiating way up into the sky.

"This unity shield will not last long. Once its bond collapses, the dark energy will come bursting through from all sides. And this will be as bad as it sounds! Make no mistake! So, here, take the two keys and get the hell out of here pronto, before Subtractoforce kicks in!" The Emperor handed the two keys to Mei.

System 1 came flying onto the terrace.

"Your Pomposity," Jafari said, "we came here through one of these unappealingly looking wormholes."

The Emperor smiled. He lifted up both his hands. A strong beam of light shot out from them. One took on the form of a dragon and the other that of a tiger. Up in the sky, these two creatures converged as a fast-spinning vortex that generated a wormhole.

"You're the awesome special effects guy, your Majesty!" Riya was mightily impressed.

"Get moving and be creative!" the Emperor shouted. "May symmetry be your guidance!"

“Thanks!” Isaac said to the Emperor. Then he ran to the spaceship along with his friends. As soon as they had shut the spaceship's door, System 1 took off and disappeared into the wormhole.

“Don't worry!” Jacky said to the crew. *“I've found a way to stabilize the flight through the wormhole. We should be coming out on the other end, nice and safe.”*

CHAPTER 9

Indeed, the flight back through the wormhole wasn't that bad. But coming out of it on the other side, back into the 21st century, made the crew instantly realize that the word "nice" had been replaced by "ugly", and the concept of "safe" had been overtaken by the ultimate "be prepared to get your asses kicked!" option. In other words, right in front of them they saw a super large black, ugly spaceship that was surrounded by countless other dark, nasty-looking vessels. Subtractoforce, big time now!

"Ever got the feeling you are not welcome back!" Jessy sighed.

"If they all fire at us, we'll have some serious technical problems in here," Jafari noted with the appropriate lack of enthusiasm.

All of a sudden, however, countless spaceships appeared out of nowhere all around them. They displayed flags from countries all over the world. From Australia, Great Britain, Spain, France, Israel, Saudi Arabia, the UAE, South Korea, Russia, Argentina, Brazil,

Mexico, Canada, New Zealand, China, countries from all over Africa, India, Thailand, Pakistan, and even more E.U. countries, even the Germans showed up!

"Look!" Isaac said, "the Octoberfest delegation's here, too."

"Reality just got more interesting. What is this?" Abdulaziz wondered.

"The second key stated *Wholeness*" Riya said, "This is the character of systems, and systems consist of numerous parts. Plus," she went on to explain proudly, "I invited all my fans from around the globe to join us!"

"But where are the Americans?" Jafari marveled.

At that very instant a thundering was heard behind System 1. All the other spaceships felt the strong vibration, too.

"What is that?" they all wondered.

Then they saw a gigantic spaceship appear above them, the shape of a hotdog and about as long and wide as Manhattan. It goes without saying… yep, this supersized vessel displayed the American flag.

"We've still got seniority with respect to all things XXL," Isaac said. To his joy, he could see that his buddies from school were commandeering the enormous spacecraft.

"With that amount of firepower," Jafari reasoned, "we might even have a chance against the Subtractoforce armada!"

Isaac opened the hailing frequency to all the ships: "Okay,

everybody, some important pointers. Shoot first and then shoot a lot more!"

"Well then, let's create some smoke around here!" Mei said, and started speeding towards the huge enemy spaceships and fired at them. The others joined in. Isaac's buddies hammered the Subtractoforce fleet with mega tough laser torpedoes.

Unfortunately, the Subtractoforce alliance had pretty good shields and some mean, weapons, too. So, the battle became a lot of the usual film-like sounds, BAM, BAM, BOOOM, BAM, BOOM, BAM-BAM and so on. It didn't sound too intelligent. But the battle was effective in being negative, which Subtractoforce got a kick out of, and therefore became even stronger (more negative).

Pretty soon, after a few rounds of status updates, "Shield's down to 50 percent. Shield's down to 40 percent. Shield's down to 30 percent," the six Posionites realized that things were really not working out.

"This resolution-dialogue-free zone is not stimulating," Riya said, who knew from her Sci-Fi-series expertise, that collapsing shields were a bummer.

Again, they were hit hard by one of Subtractoforce's torpedoes.

"I agree. Something's not right!" Isaac confirmed.

"No kidding, Sherlock!" Jafari yelled. This time there was no humor in his voice.

"Shield's down by 80 percent!" Jacky warned.

"Okay. This is the moment for a commercial break as we have a serious emotional cliffhanger here that keeps everybody wanting to see a solution, since this is the moment for a brilliant idea. How about escaping to hyperspace?!" Riya suggested based on her experience regarding avoiding the early end to a TV show.

"System 1 does not have that capability!" Mei shouted.

"How do you know! We're in possession of two magic keys. In all my critically acclaimed movies and award-winning shows, most of which were also turned into bestselling computer games, this usually means that the good guys discover they have added abilities and powers. This is it! Game on, big time!"

As they were fast running out of any other kind of option, including positive thinking, Jafari checked the computer and was totally amazed.

"Holy crap! This is sooo intense! We DO have hyperspace capabilities!"

"Then get us the hell outta here! We'll return to Subtractoforce later!" Isaac yelled.

Instantly, Jafari initiated flying (actually escaping) into hyperspace. Their spaceship was the first to disappear from the sky. Then the other spaceships of the international coalition vanished, too.

Fractions of a second later System 1, along with the other vessels, appeared in radiating hyperspace. Everybody was relieved.

At least for the moment.

"Nice!" Jafari applauded.

"Just like in real life," Abdulaziz reckoned.

But there was no time to lose on a fancy coffee break. So, Isaac suggested: "Okay, guys. Let's analyze this. Subtractoforce is really pulling out all the stops to show it's coming from an angry place. Why? It's almost as if we had all three keys already, but according to the second key, we still have to fly to the Statue of Liberty for the last key. Yet Subtractoforce is blocking the way with all its meaningless stupidity. So…"

"What do you mean?" Mei asked.

"Well, logically speaking, at the Emperor's palace, we learned that number two can be both the second key and two keys at the same time, right?!"

"So, does this funky blast from the past lead us somewhere?" Jafari wondered.

"Yeah, think about it. We have two keys," Isaac reasoned and put the two keys on the control panel. "So, the second key, is number two, and has the right to see itself as 1, 2 keys again."

"In other words," Jessy figured, "the second key could also, at the same time, be both the second and the third key!"

As soon as Jessy uttered those words, they saw a third key shimmering through, right next to the other two keys.

"Man, this has real possibilities!" Abdulaziz exclaimed.

“But why is the third key not fully materializing?” Jessy wondered.

“Maybe,” Mei suggested, “we need the third key with the word on it to create a whole structure together with *simplicity* and *wholeness*. So, what could the third word be?”

“This is so exciting. It’s like when your name gets announced as best new actress in all categories!” Riya added.

“Exactly,” Jafari said.

Isaac went on deducing, “Okay, so the Statue of Liberty was named after the Roman god Liber. He was a deity for agriculture and accompanied by two fertility goddesses, also representing agricultural good spirits …”

Abdulaziz continued this train of thought, “In addition, the law encoded in the book that the Statue of Liberty holds in her left hand, comes from the Natural Law, which, according to Plato, Aristotle, Cicero and Locke, is one fundamental law, true everywhere…”

“Oh, come on, guys!” Riya exclaimed, “it’s not that hard to guess, what the third key word is then!”

“Like NATURE!” Jessy reasoned.

“Bingo!” Riya smiled.

As soon as Jessy had uttered the word “NATURE”, the third key shone very brightly, and a powerful light filled the entire room.

CHAPTER 10

Mei's spaceship, along with all the other international coalition spaceships, popped out of the sky right next to the Statue of Liberty. It was around noon, and tourists visiting the Statue were surprised and slightly awestruck, to see far more than they expected. The Statue of Liberty began to smile and shake her torch, which then started to sparkle and radiate intensely. Naturally, this caused a lot of touristy screaming.

The real cool thing, however, was that out of the torch, the third key was tossed over to Mei's spaceship, and Jacky managed to catch the key with a robotic arm, (so let's be grateful these things were invented). This was accompanied by the Statue of Liberty saying: "May symmetry be your guidance!"

Jacky brought the third key into the spaceship, while the touristy screaming session was still going on, of course. But the appearance of the XXL United States spaceship managed to calm the sightseeing crowds down, especially when they saw all the international flags on the other spaceships. Not that they had any idea what the heck was happening, but when Isaac's buddies turned on loud Rock music and dispensed countless "I Love NY" T-shirts and mini blueberry muffins from their spaceship, the mood quickly changed from panic into, "I love this parade!" There is never a bad time for a good parade and some amusing entertainment.

Inside System 1, the real action was going on though. Jacky placed the third key, with the word NATURE inscribed on it, next to the other two keys. Isaac picked it up and read the name of the location on the other side: SWEDEN.

"SWEDEN! The Global Green Deal conference in Stockholm!" Isaac said.

"It starts today, without me!" Riya realized, almost panicking.

"But I'm sure bad-ass Subtractoforce will show up, and we know it's not good at anger management," Jafari reckoned.

"Via hyperspace, we'll be there in 30 seconds!" Mei replied.

"Let's hop over to Stockholm then!" Isaac said with an eye on the radar, "before the U.S. air force starts causing trouble."

With a loud, effective WHOOSH, all the spaceships disappeared into hyperspace again.

CHAPTER 11

This time, hyperspace presented itself as a beautiful reality, where technology and Nature coexisted positively to make life for people, plants and wildlife a joyful cyclical experience of abundance. It was a world of real quality. Here, the regenerative circle of Nature did its forever marvelous magic. The six Posionites, along with all the other crew members in the coalition spaceships, enjoyed the gorgeous, overwhelmingly mesmerizing scenery that they observed all around them. Now and then a few hand-made, game-changing cardboard signs showed up, hovering through the air, saying things like:

We demand Compliance

with Climate Science!

We want a Circular Economy

in line with Nature's Harmony!

Nature has a Symmetry Frame!

Respect its Laws, they are always the same!

"*According to my scans,*" Jacky was heard in the background, "*we are seeing the thoughts of the two young climate activists who will announce the binding Global Green Deal today.*"

Now they all watched a huge, radiating image of a smoothly rotating compass floating through the fascinating and stunning scenario. Jessy reasoned, "The Global Green Deal is based on the seminal European report *A System Change Compass* with its simple, holistic concept of building a positive future for people and planet."

So, how can this fun-for-everybody-and-the-planet vision become reality? For starters, one has to know the name of the game. **Symmetry**, like Einstein revealed it as the simple, beautiful core principle of Nature. Then, you need an **Overall Organizing System** for human civilization that is as symmetric as Nature itself. This, as a concrete vision, was provided by the European, eco-intelligent report *A System Change Compass,* with its two interacting, harmonious parts.

<u>Part One</u>, the actual compass, defining core aspects of human activity in a positive, holistic fashion as principles, and

<u>Part Two</u>, the scheme of a positive configuration of the economy, ranging from outlining concrete positive industry champions of a circular economy to guidelines for policymakers.

SIMPLE WHOLENESS of NATURE. The three sexy, timeless and space-independent key words. They were true on the smallest of scales in the universe and, therefore, also on the largest of scales. Can't get any simpler than that. If you want to impress yourself with big words and kick around some technical jargon, you just go ahead and call all that "self-similar self-organization."

Suddenly, the spaceships started to shake and seconds later...

CHAPTER 12

...the spaceships appeared way up in the blue, sunny sky above Stockholm. This time, however, they made sure to initiate camouflage mode. The last thing they wanted to do, was interrupt the vital signing of the binding agreement.

Down in the city of Stockholm, more and more international participants were arriving at the Palace for the conference. The international media coverage was huge and super chatty. Young activists were stationed all around the premises of the Palace holding up their homemade signs that had proven to be so powerful.

Several huge yachts had anchored in the harbor in front of the Palace, where the press was conducting interviews with countless experts and pundits, who (with the benefit of hindsight) all knew that things CAN happen in this world. And indeed, just as the experts had predicted for this special day, and as if to prove a point, there was the expected movement on the stock market as well.

Up in the sky above the Palace, the six young Posionites, with the help of Jacky, assessed the situation.

"Jacky's scanners are not picking up any unusual activity or signals in and around Stockholm," Mei said.

"Looks way too much like a normal day during a coffee break," Isaac stated.

"Not loving this serenity!" Jessy added, as a bad feeling grew inside of her.

"I know, you're thinking that I'm just a super pretty princess with clever, trendsetting outfits and a gifted actress who can give amazing award-winning acceptance speeches and then gets mad when afterwards she doesn't get enough gift certificates from the event's sponsors, which, by the way, is a totally false lie…"

"Where are you going with this, sugar?" Jafari interrupted.

"Given our enormous positive, sexy presence now," Riya continued, "I think Subtractoforce is weakened and its best option is to plant a super mean bomb somewhere."

"It's not totally unlikely," Abdulaziz confirmed. "It's conceivable, that Subtractoforce might even use a different dimension to do this. The thermal reaction of the bomb could cause a shockwave into our reality."

Suddenly, all the spaceships' alarm systems went off. On the computer screen of System 1, the six Posionites saw a very large dark, ghost-like yacht entering the harbor in front of the Palace. However, the yacht was not visible when they looked through System 1's window. It was obviously Subtractoforce approaching via a hidden dimension, just as Abdulaziz had guessed.

"Scan the vessel for explosives!" Mei ordered Jacky.

The scans of the ghost yacht revealed that the entire ship was filled with explosives, ready to go off in ten seconds.

"Bad timing!" Jafari sighed.

"Subtractoforce sure is a piece of work!" Isaac cursed. "I don't think we can hit it with our laser torpedoes without enhancing the explosion!"

"May symmetry be our guidance!" Jessy said staring at the countdown that was down to seven seconds.

The six Posionites and all the other crew members on the other spaceships just couldn't believe that Subtractoforce was about to win with its mindless negativity.

The clock showed three seconds. It was over.

But in that very instant, Posinity, the Chinese Emperor and the Statue of Liberty appeared, all shimmering in an eerie light, and of course, invisible to the people on the ground. In the nick of time, and together with the spaceships, they formed a magnificent laser beam and fired it at the ghostly Subtractoforce yacht. The special beam of light, enhanced by the additional energies of Posinity, the Chinese Emperor and the Statue of Liberty, catapulted the ghost vessel way up into the sky, where it exploded.

Everyone on the ground witnessed a gigantic, bright explosion. A loud BANG was heard, and people in and around the Palace reacted instantly with fear and panic. Scared, they surveyed the sky above them, but to their surprise and relief, saw a wonderful, colorful sparkling display slowly descending down to Earth. It looked like a magnificent, powerful firework that went on and on and on. Everyone understood it as a symbol of hope and joy.

A young woman reporter summed it all up: "What a wonderful surprise and great idea by the conference organizers to open this historical event with such a brilliant and unusual firework display. Everyone is feeling such joy that this binding Global Green Deal with its balanced organizing system for human society, modeled after Nature's successful creative scheme, is coming to fruition. Our thanks go to the countless young people, and everyone else, who pressed for this and made it possible: A positive future for people and planet."

Way up in the sky, all six Posionites cheered, hugged and danced with each other, as Mei put on some upbeat music. The same scene was repeated in all the other coalition spaceships. Posinity, the Chinese Emperor and the Statue of Liberty all high-fived each other three times.

An hour later, two young climate activists, a girl and a boy, stepped in front of the huge crowd of onlookers and reporters outside the Palace. Everyone fell silent, holding their breath in great expectation.

"We are delighted...," the young girl said with bursting excitement.

"...to announce to the entire world...," the boy continued beaming now with joy and pride, and together the two of them finished the historical sentence:

"...the binding Global Green Deal has been ratified!"

Instantly, the crowd in front of the Palace erupted with great joy and jubilation. People jumped up and down, embraced each other, kissed and danced. Their happiness reverberated around the globe, as people everywhere celebrated on hearing the good news.

Posinity, the Chinese Emperor and the Statue of Liberty started dancing, too. It was astonishing to see just how much they were into hip hop and cool dance moves. The coalition spaceships, with their equally wild, dancing, cheering crews radiated in multicolored hues, flew happy loops and shook and wiggled to the beat of the music that was blasting joyfully into the vast blue, beautiful sky of planet Earth.

SYSTEM CHANGE COMPASS
Title Song to the Novel

Music & lyrics: George Hohbach
Arrangement: Alfred Huff

VERSE 1
We don't cure symptoms
But we do
Make all positive
To see us all through

BRIDGE
Through, through, through
To, to, to
A better world
America, Africa,
Europe and Asia

CHORUS
System Change Compass
Our guiding Star
Orientation
Everything is on par
Let's face it, face it!
Let's face it head on!
The future is won
The crisis overcome
Come on! Come on! Come on!
Come on! Come on! Come on!

VERSE 2
We are not superficial
But a system's crew

To make it all work
To see us all through

BRIDGE
Through, through, through
To, to, to
A better world
America, Africa,
Europe and Asia

CHORUS
System Change Compass
Our guiding Star
Orientation
Everything is on par
Let's face it, face it!
Let's face it head on!
The future is won
The crisis overcome
Come on! Come on! Come on!
Come on! Come on! Come on!

System Change Compass
Our guiding Star
Orientation
Everything is on par
Let's face it, face it!
Let's face it head on!
The future is won
The crisis overcome
Come on! Come on! Come on!
Come on! Come on! Come on!
Come on! Come on! Come on!
Come on! Come on! Come on!
Come on!

SYSTEM CHANGE COMPASS

Music & lyrics: George Hohbach
Arrangement: Alfred Huff

GLOBAL GREEN DEAL
Rap to the Novel

Music & lyrics: George Hohbach
Arrangement: Alfred Huff

VERSE
Einstein liked simplicity
He loved simply symmetry
He looked at reality
Symmetry, his huge discovery:

CHORUS
The laws of Nature always the same
Nature has a symmetry frame
What we now need
A Global Green Deal
System Change Compass
The steering wheel

Amsterdam's going circular
All Cities: Let's do this near & far!

VERSE 2
So, symmetry appears
time & again, yo
And for us that means
the awesome following, bro:
Space-time, mass
& kicking energy, man
In empty space guarantee
Forever lightspeed's constancy

So, all you guys see, dude
Is amazing unity
Radiating, real
dynamic harmony

CHORUS
The laws of Nature always the same
Nature has a symmetry frame
What we now need
A Global Green Deal
System Change Compass
The steering wheel

Amsterdam's going circular
All Cities: Let's do this near & far!

VERSE 3
Dude, quantum mechanics, too
is how symmetry behaves
The small particle so likes
Cool infinite circular waves
'Cause symmetry equates
unity with sexy infinity
That gives you uncertainty,
probability, eventually
all adds up via gravity
into our large world reality

CHORUS
The laws of Nature always the same
Nature has a symmetry frame
What we now need
A Global Green Deal

System Change Compass
The steering wheel

Amsterdam's going circular
All Cities: Let's do this near & far!

VERSE 4
Life uses circularity
To keep all moving in
amazing unity, 'cause
the circle contains infinity
Turn it forever and
No change you'll see
Symmetry as creativity
So, our economy
In curved space-time reality
Needs holistic Circularity

CHORUS
The laws of Nature always the same
Nature has a symmetry frame
What we now need
A Global Green Deal
System Change Compass
The steering wheel

Amsterdam's going circular
All Cities: Let's do this near & far!

CHORUS
The laws of Nature always the same
Nature has a symmetry frame
What we now need

A Global Green Deal
System Change Compass
The steering wheel

System Change Compass
Global Green Deal

PART 2
BACKGROUND INFORMATION ON *A SYSTEM CHANGE COMPASS* & FURTHER HOLISTIC CONCEPTS

I. THE SEMINAL REPORT "A SYSTEM CHANGE COMPASS: Implementing the European Green Deal in a Time of Recovery"

"Our actions now will define the future of our society (…) it is no longer enough to simply act. We must act quickly, systematically and together."
Martin R. Stuchtey, Sandrine Dixson-Declève, Janez Potočnik, et al.
(A System Change Compass, October 2020, p.ii)

The seminal systematic report, *A System Change Compass* offers a balanced cybernetic overview, *a much-needed organizing system*, regarding how a coherent and concise holistic set of key principles (the compass) gives **science-based** orientations on managing socio-economic activities in a fashion that mirrors the holistic way Nature works.

The report was co-authored by members of The Club of Rome and SYSTEMIQ, such as Martin R. Stuchtey, Sandrine Dixson-Declève, Janez Potočnik and others. (see: A System Change Compass, October 2020, p.iii)

The report's symmetric, systematic configuration:
An Overall Organizing System for Human Civilization

The report's symmetric, systematic configuration:

Overall, the report consists of a positive-holistic (all-inclusive), socio-economic and Nature-based system that has **two** linked, interconnected, and equally beneficial parts:

1) A concise IDEA COMPASS OF CHANGE consists of 10 positive, interconnected principles that define human actions, visions and goals such as financing, governance, prosperity, competitiveness or consumption. As a whole, it provides holistic empowering orientations or directions for:

2) An equally positive-holistic description of different parts or REAL-WORLD ECONOMIC ECOSYSTEMS of human society, that are kept active and evolve by beneficial, leading, future-fit industries, the so-called CHAMPIONS, interacting in reinforcing positive cycles, in an all-inclusive framework or OVERARCHING SYSTEM OF ENABLING POLICIES.

As a result, the report presents a much needed, holistic (all-encompassing) ORGANIZING SYSTEM for human civilization. It provides a helpful positive guiding system to assist making the necessary economic changes interconnected, mutually beneficial and effective.

"It is the predicament of mankind that (…) failure occurs in large part because we continue to examine single items (…) without understanding that the whole is more than the sum of its parts, that change in one element means change in the others."
Donella H. Meadows, Dennis L. Meadows, Jørgen Randers, William W. Behrens III
(The Limits of Growth,1972, p.11)

II. THE CIRCULAR ECONOMY (CE)

1. <u>THE DEFINITION OF THE CIRCULAR ECONOMY</u>

The Circular Economy is a simple lifestyle based on Nature's circularity. It aims to generate positive, balanced effects for people and planet and a prosperous future for all.

2. <u>THE GOALS OF THE CIRCULAR ECONOMY</u>

<u>The main holistic (all-encompassing) goals:</u>

a. To ensure the wellbeing and the generation of positive effects for people and planet, only so-called '*safe materials*' or substances are used.

b. That the 'safe materials or substances' are of continuously *high-quality* so that they can be <u>fully recycled</u> (and are not just downcycled to end up as waste)

c. The '*safe materials*' are beneficial for people and planet to ensure both the protection and regeneration of the environment: soil, water, air, plant life, wildlife, human beings (this goes beyond just being sustainable, 'keep doing what is being done as long as possible', as the goal is to continuously create positive effects)

Therefore, the positive Circular Economy aims at being eco-intelligent and climate-smart, <u>offering solutions</u> for:

- the (toxic) waste problem (which includes, for example, the issue of carbon dioxide (CO_2) in the atmosphere and oceans)
- the depletion of resources
- health and environmental issues, in general terms.

These problems are essentially created by the linear economic model of *take-make-waste*.

3. <u>NATURE'S HOLISTIC BLUEPRINT FOR THE CE</u>

The CE respects that Nature is a **BALANCED UNITY-NETWORK**, a <u>positive</u> (all-encompassing) **WHOLE**.

BALANCE comes from the fact that, as physicist Albert Einstein revealed, **SYMMETRY** (harmony, simple sameness) plays the central role in Nature (the cosmos, the universe), because the laws of Nature are **THE SAME** (symmetric, equal, identical) for everybody (all observers) everywhere at all times.

This is so, according to Einstein, as all key phenomena of Nature (the universe), that is, space, time, energy, mass and the constant lightspeed are **ONE** intertwined **UNITY,** or **WHOLE**.

UNITY, in the end, means **SYMMETRY** (harmony, balance, equivalence, sameness) in terms of **<u>doing the same things (having the same effect),</u>** like maintaining the fundamental **UNITY OF NATURE**.

The **CIRCLE** is the simple and most symmetric shape, as one can rotate it infinitely many times *without causing change* to the shape. The shape therefore stays **THE SAME** (symmetric). This respects the aspect of **UNITY** in a very effective and also creative fashion, since the circular motion permits holistic network interactions.

NATURE uses the circular motion on all levels as much as possible in its unity-network on planet **EARTH**. (also see *Why Symmetry runs the Positive Circular Economy* by George Hohbach, 2021)

4. <u>THE CE'S SCHOOL OF THOUGHT AND THE BRIGHT MINDS THAT CONTRIBUTED</u>

THE BLUEPRINT

The symmetric
circularity of Nature

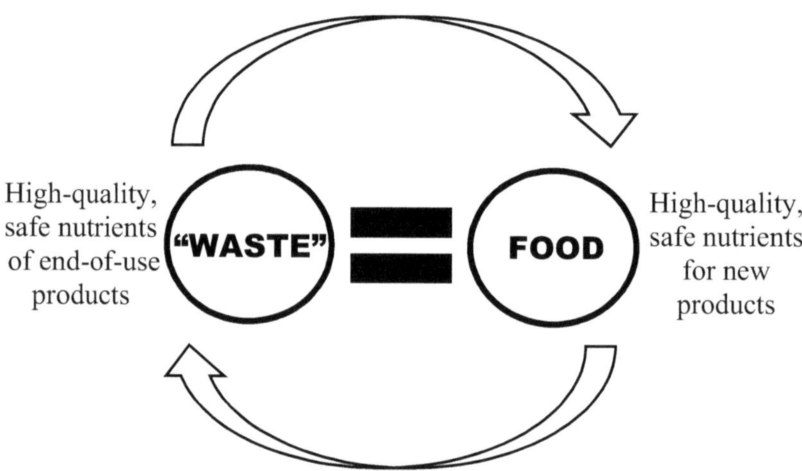

The same (symmetric) circularity applies to both **biological** and **technological** substances (nutrients) according to the school of thought of the holistic design concept of **Cradle to Cradle (C2C)**. In the **biosphere**, the cycles are open, and in the **technosphere**, the cycles are closed loops to keep hazardous but valuable technological nutrients out of the environment.

"…a sustainable future can only be realized if we work together and create new and unusual alliances and partnerships."
Prins Carlos de Bourbon de Parme
(Prins Carlos de Bourbon de Parme: Opening Address, C2C-Congress, 2017)

The countless environmental problems that humanity is currently facing, are caused by the <u>linear economic model of "take-make-waste"</u>, **that destroys the positive-holistic, circular infrastructure existing in a warped (round) spacetime environment**.

"In all my efforts I have tried to make it clear that all these subjects suffer the same problem because they have become detached from the important basic principles – the principles that produce the active state of balance which is just as vital to the health of the natural world as it is for human society. We call the active but balanced state `harmony´…"
HRH The Prince of Wales
(Harmony, written together with Tony Jupiter & Ian Skelly, 2010, p.5)

"Justice is key to a sustainable economic model if it is supposed to work globally. It's the only way to prevent the ecological question being played off against the social one. Both belong together and can only be solved in unison."
Maja Göpel
(Unsere Welt neu Denken (Rethinking our World), 2020, p.179, Author translation)

LINEAR is a line and very simply denotes the **MINUS SIGN**, which **is sent through the interconnected network of Nature's POSITVE circles (metabolisms) and destroys them.**

MINUS

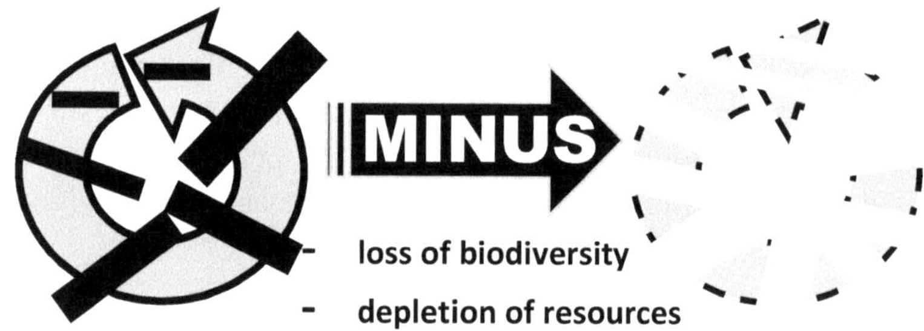

LINEAR/MINUS CHOPPING UP NATURE'S MOST SYMMETRIC, HARMONIOUS, BALANCED, HOLISTIC CIRCLE

MORE

NEGATIVE

On the other hand, the **CIRCLE represents the PLUS SIGN**, that contains both ADDED UP/ADDING UP (UNITY#1) AND THE ADDITIONAL (MORE, #2). So, PLUS can create a CIRCULAR or CYCLIC motion or CURVATURE like this: CONSTANTLY (i.e., ADDED UP as being constant) ADDING UP MORE (THE ADDITIONAL) INTO A UNIT (ADDED UP) TO SHOW MORE TO ADD UP MORE INTO A UNIT (ADDED UP) TO SHOW MORE and so on forever. This creates CURVATURE, CYCLES, CIRCLES and CYCLIC/CIRCULAR MOTION.

Understanding PLUS (ADDED UP/ADDING UP & THE ADDITIONAL) as the circular motion

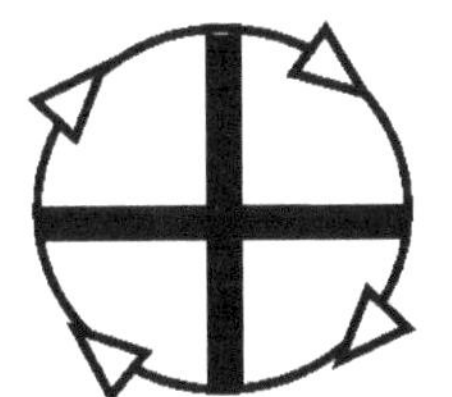

Constantly ADDING UP MORE THAT SHOWS creates CURVATURE and the Cyclic/Circular motion.

PLUS 

GENERATING THE VIRTUOUS CIRCLE

- **NATURAL NETWORKS +**
- **REGENERATION +**
- **ABUNDANCE +**
- **DIVERSITY +**
- **BEAUTY, HARMONY +**
- **STABILITY +**

MORE POSITIVE

The linear model is negative, that is MORE by itself. Why? Because, the linear model wants to constantly eliminate (subtract) something, and that something is waste. So, the output is something that is eliminated (subtracted). Accordingly, the concept attempts to destroy UNITY (#1) by subtracting something (the waste) from it. If you destroy unity, then you create MORE in a negative fashion, and since MORE has no end, you manifest INFINITY in a negative fashion. This creation of NEGATIVE INFINITY we can call environmental problems. Take for example the countless toxins or nano-plastic particles now found all over the planet.

NEGATIVITY also means LESS, as it implements the MINUS SIGN. Consequently, we deplete natural resources and cause mass extinction.

The countless circular motions of Nature are being destroyed by linear actions, be it in the soil, water, air or regarding carbon dioxide.

THE SYMMETRIC (POSITIVE) CIRCULAR ECONOMY

The symmetric or positive Circular Economy is a smart, holistic lifestyle, building on the cyclic motion of Nature. It aims towards beneficial effects for people, business and planet.

Thankfully, many people, and the number is growing fast, realize the linear economic model is not working out well. It focusses on being efficient, being MORE in terms of being faster and producing more profit, but excludes the UNITY TERM. Consequently, the linear model is not effective, as it does not implement SYMMETRY (UNITY, BEING POSITIVE, WHOLENESS) on a high, stable level as the positive circular motion does.

> "Nature gave us the correct recipe."
> Michael Braungart and William McDonough
> (The Upcycle, 2013, p.211)

<u>Let's see how the Circular Economy (CE) can be most POSITVE (SYMMETRIC):</u>

The simple, symmetric **<u>Cradle to Cradle (C2C) design concept</u>** co-founded by world-renowned chemist Michael Braungart and pioneering environmental U.S. architect William McDonough (see *Cradle to Cradle* by Michael Braungart & William McDonough, Vintage, 2009), is generally regarded as the <u>guiding path towards the global circular economy from the U.S. to the E.U. to China</u> (see *The Green Industrial Revolution* by Woodrow W. Clark II & Grant Cooke, Butterworth-Heinemann, 2015, p. 368) as well as being <u>the design philosophy at the heart of the circular economy</u> (see *The Circular Economy – A Wealth of Flows* by Ken Webster, Ellen MacArthur Foundation Publishing, 2017, p.16). The WHO regards the Cradle to Cradle design philosophy—with its **<u>cyclic bio- and technosphere</u>** based on Nature's **<u>positive, holistic</u>** concept of the nutrient cycle—as an integral foundation of the CE (see *Circular Economy and Health*, WHO, 2018, p.74). The eco-intelligent and climate-smart Cradle to Cradle design concept **aims at making everything positive from the start for all involved parties. That is to say, people, business and planet.**

"I've watched as many of the concepts presented in *Cradle to Cradle* have taken root at the U.S. Postal Service and NASA (…) and in countries around the world. I've seen how these simple ideas, when put into practice, can improve productivity and make people happier and healthier."
U.S. President Bill Clinton
(Foreword to The Upcycle by William McDonough and Michael Braungart, 2013, p. xvii)

"If we are to succeed (…) it is very important to develop a circular economy based on cradle-to-cradle principles (…) what China's government wants to achieve."
Madame Deng Nan,
China's Party Secretary for Science and Technology
(as quoted in Sense & Sustainability by Ken Webster and Craig Johnson, 2008, p. 29)

In his profound book *The Circular Economy – A Wealth of Flows*, Ken Webster, circular economy expert and head of innovation at the Ellen MacArthur Foundation, defines the concept of the circular economy as a non-linear system of relationships of parts that work as an all-inclusive dynamic pattern, where all parts interact and influence each other. How we perceive the world, how the economy functions and how we operate within this reality, according to Webster, is a **symmetry**, a matter of sameness, since the natural, circular and holistic premise of the circular economy is omnipresent.

Further, as Webster delineates in chapter 11, this **symmetric state,** in which Nature's wholeness, the systematic interaction of all parts of reality, the dynamics of the circle of this holistic interaction and the collective, organic mindset of humanity are one creative force. It can be achieved globally with the support of information and communication technology (ICT), enabling the principle of access over ownership to become reality. The golden thread of digital interconnectedness in a circular economy (CE) will then be the orchestrated interaction of systematic knowledge, <u>biological and technological materials, (nutrients, the basis of the CE)</u> and clean energy, to benefit both people and planet, and profit in order to prosper long-term while optimizing the entire system.

"The (…) balanced circular economy is added to the much-needed solutions of our time."
Maximilian Moser & Erwin Thoma
(Die Sanfte Medizin der Bäume, (The Gentle Medicine of Trees), 2018, p.144, Author translation)

Given its push towards continuous innovation, the circular economy is both a diverse whole and an open system that evolves based on the convergence of powerful, holistic ideas and scientific discoveries about Nature's wholeness.

"…the solution is all to do with finding a path that puts Nature and her virtuous circles back at the heart of things rather than on the periphery."
HRH The Prince of Wales
(Harmony, written together with Tony Jupiter & Ian Skelly, 2010, p.297)

Consequently, numerous scientists, thought leaders and organizations have contributed and are contributing to the emergence of the positive, holistic circular economy as a school of thought. Among these scientists, pioneers and thought leaders are:

- Albert Einstein, who demonstrated that Nature (the Cosmos) is a whole, a unity, where all aspects of reality work together. The Cosmos is a harmonious, balanced reality. (see *Relativity – The Special & The General Theory* by Albert Einstein, Martino Publishing, 2010). With his work, Einstein revealed the central role of symmetry in Nature, as he not only based his revolutionary discoveries on symmetry as the guiding, primary principle of Nature but also eventually unveiled how central symmetry actually is in Nature.

- Isaac Newton had previously shown that the cosmos is a whole, since on the large scale it is governed by gravity as the uniting phenomenon. Newton's laws of motion conform with each other, working seamlessly in harmony with Nature, again establishing that there is unity, or wholeness in Nature.

- Before Newton, Galileo contributed enormously to the realization of the importance of symmetry (oneness, unity) in Nature. Albert Einstein consequently achieved his revolutionary scientific discovery about the harmonious unity, simplicity and elegance of Nature building on both Galileo and Newton.

- Physicist Erwin Schrödinger, a contemporary of Einstein, described the actions of a quantum (small particle like an electron) via an overall wavefunction. This overall wavefunction is a holistic whole, consisting of infinitely many single wavefunctions (spreading out over space), each of which has a shape of a <u>revolving</u> corkscrew.

- Mathematician Emmy Noether, confirming the importance of symmetry as unveiled by Albert Einstein also demonstrated that time (which builds <u>on cycles</u>) offers a very flexible "terrain" for symmetry in Nature. (The connection between the continuously symmetric laws of Nature in time and energy conservation, i.e., that the amount of energy stays symmetric or the same).

- John T. Lyle with his ideas of *Regenerative Design* establishing the framework of the Circular Economy.

- In his 1966 essay *The Economics of the Coming Spaceship Earth*, economist Kenneth Boulding describes an economic model that is based on Nature's circularity.

- World record-breaking sailor, Ellen MacArthur, and Circular Economy Expert, Ken Webster, by promoting the Circular Economy internationally and persuading major players (companies, organizations) to commit to CE practices via the many trendsetting initiatives and partnerships of the Ellen MacArthur Foundation. (see *Full Circle* by Ellen MacArthur, Penguin Books, 2010; *The Circular Economy* by Ken Webster, The Ellen MacArthur Foundation, 2017; *Sense & Sustainability* by Ken Webster and Craig Johnson, TerraPetra, 2008); Collin Webster, Education Content Manager at the Ellen MacArthur Foundation with, e.g., his TEDxYouth talk: *"Re:Thinking the Future"* or the film *"System Reset"*, about a regenerative economy, on which he collaborated.

"We need to keep the precious materials we have in constant circulation."
Ellen MacArthur
(Full Circle, 2010, p.362)

"… how we see the world; how the economy really works and how we can act within it. There is an important symmetry here."
Ken Webster
(The Circular Economy – A Wealth of Flows, 2017, p.7)

- The Club of Rome with books like *The Limits to Growth* or *Come On!* by Ernst Ulrich von Weizsäcker and Anders Wijkman

"Our greatest inventions, discoveries and acts of creativity come when apparent contradictions are reconciled. Integral thinking is thinking that is able to perceive, organize, reconcile and reunite the component elements and arrive at a truer understanding of the underlying reality."
Ernst Ulrich von Weizsäcker & Anders Wijkman
(Come On!, 2018, p.199)

"The transition to sustainability (…) there will also be changes to (…) the human perspective on nature."
Jorgen Randers
(2052, 2012, p.13)

- Walter Stahel with his concept of the *Performance Economy*, who is also credited with coining the term Cradle to Cradle in the 1970s (see *The Circular Economy – A User's Guide* by Walter Stahel, Routledge, 2019).

"…most of the topics jeopardizing the world society, listed as the objectives of the UN Sustainability Goals (…) could be tackled simultaneously by the shift to a circular industrial economy."
Walter R. Stahel
(The Circular Economy, 2019, p.90)

- HRH The Prince of Wales with his support for organic farming *(see Harmony: A New Way of Looking at Our World* by HRH The Prince of Wales, Harper Collins Publishers, 2010)

"… to understand the significance of Nature's processes and to live by her cyclical economy."
HRH The Prince of Wales
(Harmony, written together with Tony Jupiter & Ian Skelly, 2010, p.15)

- Arnold Schwarzenegger, former Governor of California, with his many ground breaking initiatives, such as the Green Chemistry Initiative (see *Total Recall* by Arnold Schwarzenegger with Peter Petre, Simon and Schuster, 2012) and supporting the launch of the non-profit, now called *C2C Products Innovation Institute* in California in 2010. California's EPA (Environmental Protection Agency) recommendation in December of 2008 to:

"MOVE TOWARD A CRADLE-TO-CRADLE ECONOMY."
California EPA
(Press Release, December 16, 2008)

"The time is now for us to go beyond simply being "less bad" and to lead the world in the invention and innovation of *"more good"* with Cradle to Cradle products and a prosperous Cradle to Cradle economy. Together, we will inspire and transform the world."
Arnold Schwarzenegger, May 20, 2010,
(at the launch of the non-profit C2C Products Innovation Institute at Google's headquarters)

- Janine Benyus, establishing the discipline of *Biomimicry* (see *Biomimicry* by Janine M. Benyus, William Morrow,1997)

- Rachel Carson with her seminal book *Silent Spring* denoting how treatment of the environment by humans also impacts themselves.

"The balance of nature (…) cannot safely be ignored any more than the law of gravity (…) it is fluid, ever shifting, in a constant state of adjustment. Man, too, is part of this balance."
Rachel Carson
(Silent Spring, 2002, p.246)

- Gunter Pauli with his concept of the *Blue Economy* that focusses on the holistic big picture. (see *The Blue Economy 3.0* by Gunter Pauli, Xlibris, 2017)

"Nature's principle of optimisation of the whole, rather than maximisation of selected parts (…) involves recognising the delicate balance among all factors..."
Gunter Pauli
(The Blue Economy, 2017, p.7)

- Paul Hawken, Amory Lovins, L. Hunter Lovins, describing the concept of Natural Capitalism (see *Natural Capitalism* by Paul Hawken, Amory Lovins, L. Hunter Lovins, 1999, Little, Brown and Company; also see *Drawdown – The Most Comprehensive Plan Ever Proposed To Reverse Global Warming* edited by Paul Hawken, Penguin Books, 2017)

- Kate Raworth and her concept of the holistic Doughnut Economics. (*Doughnut Economics* by Kate Raworth, Penguin Books, 2017)

"*Doughnut Economics* sets out an optimistic vision of humanity's common future: a global economy that creates a thriving balance thanks to its distributive and regenerative design."
Kate Raworth
(Doughnut Economics, 2017, p.286)

- Bill Mollison and David Holmgren defining the concept of *Permaculture*, as well as Sepp Holzer

- Gabe Brown with *Dirt to Soil* about *Regenerative Agriculture* and also providing an excellent definition of holistic management.

  "The premise of holistic management is that nature functions in wholes."
  Gabe Brown
  (Dirt to Soil, 2018, p.36)

- Charles Massy with his book *Call of the Reed Warbler* that focusses on the transition towards an open, holistic mindset in farming.

  "So the only way to get our heads right for managing complex creative systems is to think holistically, flexible and openly."
  Charles Massy
  (Call of the Reed Warbler, 2017, p.338)

- Ibrahim and Helmy Abouleish with the *SEKEM* initiative, pioneering biodynamic farming in Egypt.

- The *Circular Economy Club*, founded by Anna Tari, that globally promotes the implementation of CE principles.

- The *C2C NGO* with its comprehensive program to raise awareness of the C2C school of thought.

- In the article *Impacts of Green New Deal Energy Plans on Grid Stability, Costs, Jobs, Health, and Climate in 143 Countries* (One Earth 1, 449-4463 December 20, 2019), Mark Jacobson et al. examine a scenario where energy comes from Wind, Water and the Sun (WWS).

- Platforms like *We the Planet* try to bring people of all ages together to work towards a better future.

- *Scientists for Future*, a non-partisan and interdisciplinary organization, addresses issues like climate change or crises like the loss of biodiversity and highlights possible solutions.

- Many young climate activists and thought leaders in general have formed organizations and started initiatives around the world to address the pressing global issues and promote solutions, like *Fridays For Future, Re-Earth Initiative, Sunrise Movement, Foreign Policy Youth Collaboration, Climate Live,* etc.

- The NGO *Kiss the Ground,* promotes regenerative agriculture and provides comprehensive education in this field (also see the book *Kiss the Ground* by Josh Tickell, 2017).

"Soil, dirt, sand, dust—call it what you will—the way we care for this precious foundation will determine the fate of our species. That fate is not a "someday maybe" kind of fate. It's today. It's right now."
Josh Tickell
(Kiss The Ground, 2017, p.292)

- In 2021, the cities of New York, Toronto, Amsterdam, Glasgow and Copenhagen joined forces via the *Circular Innovation City Challenge* to find intelligent circular economy solutions.

- The foundation *JUSTDIGGIT* works on regreening land to create a positive impact on the climate. Ancient and modern techniques and technologies are combined to regenerate dry land via landscape restoration programs. Educational programs are also offered.

- The *Apparel Impact Institute*, that implements and supports holistic initiatives, led by its president Lewis Perkins.

- In earlier times, polymaths like Leonardo da Vinci, author Johann Wolfgang von Goethe, explorer Alexander von Humboldt, U.S. diplomat and author George Perkins Marsh, U.S. author Henry David Thoreau, the father of the U.S. National Parks, John Muir, British economist Arthur Cecile Pigou, and Charles Darwin, to name but a few, contributed to the holistic understanding of Nature.

- Ancient cultures around the globe also understood both the importance of cycles in Nature and its wholeness as seen in the Yin-Yang symbol, Feng Shui, and Stonehenge.

Apart from the seminal work of former U.S. Vice President Al Gore (see *An Inconvenient Truth*, 2006), it is also of systematic value to consider **the comprehensive findings summarized in the compelling book *Rethinking Humanity* by James Arbib and Tony Seba** (June, 2020). In a profound, tangible and captivating fashion, the authors delineate that the five key sectors <u>information, energy, food/agriculture, transportation and materials</u> are subject to fast, major changes due to sweeping new technologies. This is based on historic developments, the convergence of scientific fields and the fast evolution of technology. Implemented in the correct holistic responsible way, these changes offer opportunities to human society that can both help tackle environmental problems (including climate change) and manifest an unprecedented prosperous future for all in an Age of Freedom.

Various papers, for example, *Completing the Picture – How the Circular Economy Tackles Climate Change* by the Ellen MacArthur Foundation of 2019, *The Circular Economy – A Powerful Force to Climate Mitigation* by Material Economics and its partners, *The Future of Nature and Business Policy Companion* by the World Economic Forum of 2020, and *Tackling the Climate Crisis and the Corona Pandemic Recession* by the World Future Council of 2020, all emphasize the need to better integrate Nature into our economic concepts. This message is also conveyed by documentaries such as *2040* by Damon Gameau and *Before the Flood* by Hollywood actor/environmentalist Leonardo di Caprio. *Drawdown*, a book edited by Paul Hawken, lists and describes, in a comprehensive fashion, key approaches to overcoming climate change and the environmental crises. The UN report *Making Peace with Nature – A scientific blueprint to tackle the climate, biodiversity and pollution emergencies* (2021) highlights how transforming the relationship that humankind has with Nature is essential to achieving a prosperous future which implements the Sustainable Development Goals to benefit all in harmony with Nature.

"…to balance the need for growth with the need to maintain stability."
James Arbib & Tony Seba
(Rethinking Humanity, 2020, p.29)

"The growth of business for good incorporates a new meaning of growth which must embrace (…) connecting growth with positive impact on the world."
Marga Hoek
(The Trillion Dollar Shift – Achieving the Sustainability Goals, 2018, p.7)

"An essential element of all living systems is a dynamic balance between self-preservation and integration. Neither aspect is either good or bad. Good is a dynamic balance between both."
Ibrahim Abouleish
(Die Sekem-Symphonie/The Sekem-Symphony, 2016, p.157, Author translation)

"We humans are not stupid. We are not ruining the biosphere and future living conditions for all species because we are evil. We are simply not aware."
Greta Thunberg
(No One is Too Small to Make a Difference, 2019, p.73)

THE SYMMETRY CIRCLE

of eco-intelligent and climate-smart concepts and initiatives

The SYMMETRY CIRCLE lists thought-leading concepts that can strengthen symmetry, and Nature's core (balance, network unity, harmony, wholeness, circular action, holism), to make the planet and society MORE SYMMETRIC again. The list is, of course, incomplete and cannot cover all the great, essential efforts. The singled-out, listed key concepts can interact in reinforcing cycles, as they are positively holistic, overlap or complement each other. This way, a more symmetric, hologram-like world view of dynamic unity/wholeness can emerge, be imagined, communicated, felt in its positive, all-unifying force of powerful interacting loops and materialize to benefit all: people, business and planet.

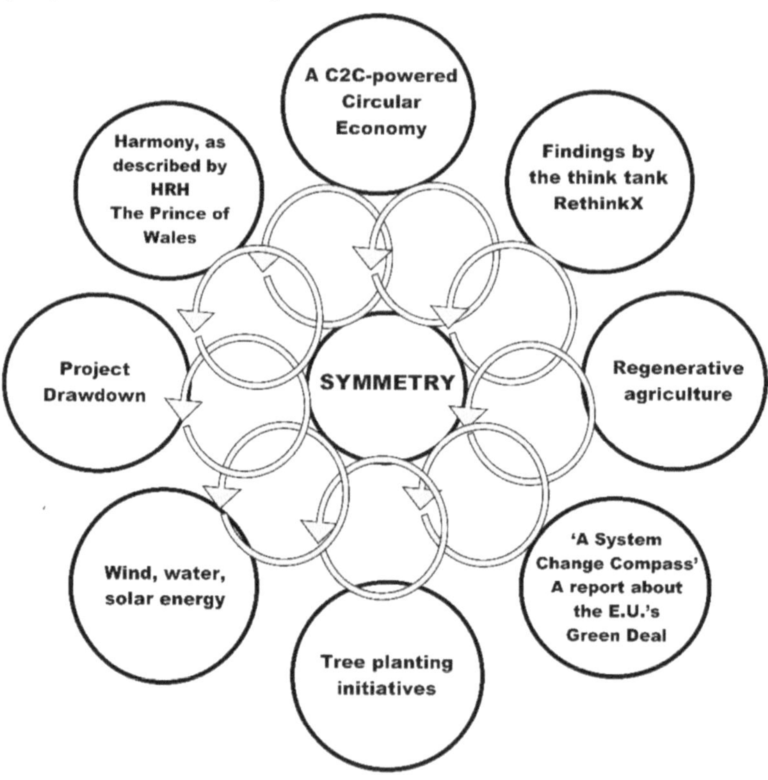

<u>"A System Change Compass: Implementing the European Green Deal in a time of recovery"</u>

<u>THE BIG PICTURE & OVERALL ORGANIZING SYSTEM</u>
The holistic orientation and symmetric, societal Organizing System for governments, companies and society as a whole, as presented by The Club of Rome and SYSTEMIQ.

THE POSITIVE CIRCULAR ECONOMY
Nature's Circular, Symmetric Recipe

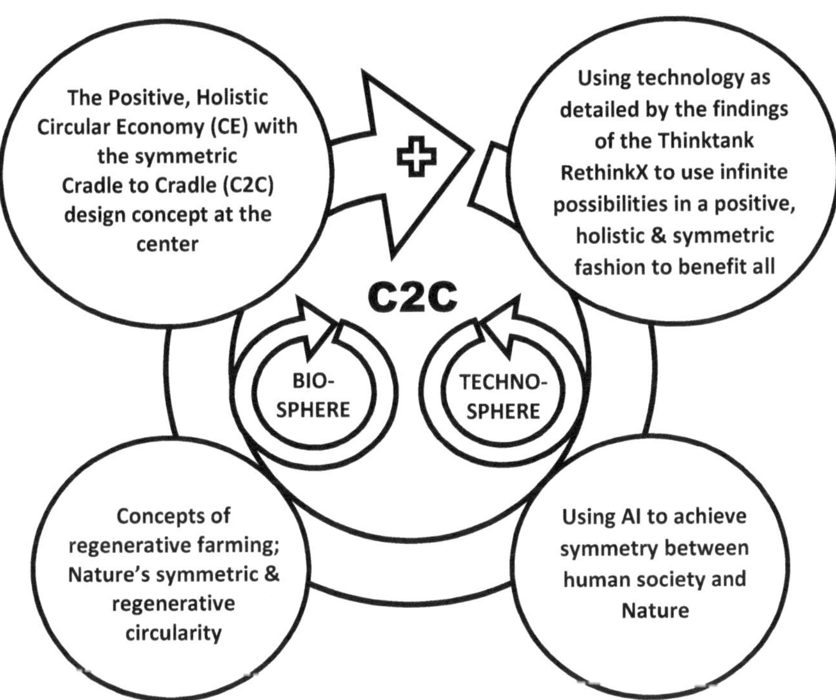

<u>Nature's Foundation:</u>
Albert Einstein's findings on the core role of SYMMETRY in Nature

The implementation of these concepts also supports and promotes the implementation of the UN's 17 Sustainable Developments Goals (SDGs) that present an intertwined whole.

"…if Western Civilization is able to reorganize itself (…) the ability of Western Civilization to survive and go on to increasing prosperity and power will be bright."
Carroll Quigley
(Tragedy and Hope, 1966, p.7)

Will this renewal occur again, also globally, now in the 21st century?

"So the state founded on natural principles is wise as a whole."
Plato
(The Republic, translated by Desmond Lee, 2007, 428e)

"…the form of the good; once seen, it is inferred to be responsible for whatever is right and valuable in anything (…) And anyone who is going to act rationally (…) must have sight of it."
Plato
(The Republic, translated by Desmond Lee, 2007, 517b)

"There is indeed a law, right reason, which is in accordance with nature; existing in all, unchangeable, eternal. Commanding us to do what is right, forbidding us to do what is wrong. It has dominion over good men, but possesses no influence over bad ones. No other law can be substituted for it, no part of it can be taken away, nor can it be abrogated altogether. Neither the people or the senate can absolve from it. It is not one thing at Rome, and another thing at Athens: one thing today, and another thing tomorrow; but it is eternal and immutable for all nations and for all time."
Cicero
(The Republic of Cicero, translation by G.W. Featherstonhaugh, Introduction, p.13)

ACKNOWLEDGEMENTS

Such a book does not come into existence without the enthusiastic participation, help and support of many people. But I wish to single out a few here. First of all, many thanks to former E.U. commissioner for Science & Research and for the Environment and co-author of the report *A System Change Compass* Janez Potočnik for his encouragement to write this novel. Many thanks to musician and composer, Alfred Huff, for producing the arrangement of the novel's songs, as well as to Mrs. Frick for her support regarding the preparation of the cover, and to my agent Lloyd Robinson, for his support and encouragement over many years. Many thanks to Marialine Verdickt, Ambassador at Circular Economy Institute, for her support and inspiration. Finally, I want to thank business development and marketing expert, Lynda Brandt, for her productive collaboration in regard to putting the finishing touches on this book.

About the Author George Hohbach

George Hohbach studied Law and has collaborated for nearly two decades with Suite A Management, a talent and literary agency in Beverly Hills. Alongside his writing and involvement with Value Investing, George studied Albert Einstein's revolutionary symmetric equivalence principle for many years. As Head Writer, together with artists from the U.S. and around the globe, he produced several, authorized, solution-driven Action Comedy Novels aimed at young adults, created around the eco-intelligent, climate-smart school of thought, Cradle to Cradle (C2C), based on the works of its co-founder, chemist Michael Braungart, who developed C2C in cooperation with U.S. pioneering environmental architect William McDonough. Using the connections between Albert Einstein's groundbreaking symmetrical equivalence principle with the ensuing mathematical principles, Cradle to Cradle, and the positive Circular Economy, he has also written and published various scientific summaries on system-theoretical analyses. George has a romantic comedy with U.S. author Robin Palmer about the positive Circular Economy currently in production.

George Hohbach is not only a writer but also a composer and artist, and has produced various pop songs on C2C, the Circular Economy or *A System Change Compass*. His prolific artwork in various media, including paintings, music, books, and documentaries, is presented in galleries, educational centers, companies, and museums throughout the world. George lives and works near a beautiful Nature Reserve from where he gathers much of his inspiration.

Ehrengard Hohbach
is a former physician with a specialty in holistic medicine. She gave numerous lectures, seminars and courses on that topic. Her expressionist, symbolic paintings—also inspired by C2C and Einstein's findings about symmetry—are exhibited in galleries, companies and museums domestically and abroad. Her artistic and literary work was covered several times on radio programs as well as in newspaper articles.

Also by George Hohbach & Ehrengard Hohbach with Scot Marcano;
illustrations by Juan Romera:

C2C NOVEL *MICHAEL & MIA*

An Adventure in the Positive Cycle of Nature
illustrated Middle Grade Novel

Two young chemistry fans, haunted by a polluting, mad robot, discover a stunning, circular solution in Nature at the last minute that can save their home planet.

A humorous adventure based on the life of C2C co-founder Michael Braungart with background information on the school of thought of C2C.
Plus: the theme song *NATURE LIVES* as sheet music.

C2C NOVEL *AGENT C2C*

Positive for People and Planet
illustrated Young Adult Novel

Agent C2C and his two American cohorts have to find solutions to two dangerous international crises and the growing environmental challenges – fast!

A humorous action adventure based on the life of C2C co-founder Michael Braungart with background information on the school of thought of C2C. Plus: the title song *AGENT C2C* as sheet music.

Also by George and Ehrengard Hohbach:

NOVEL
FOREVERCIRCLELAND

illustrated Children's Novel

Preparing for his Master's exam, magic apprentice Max inadvertently embarks on a dangerous adventure that reveals how beneficial Nature's circular magic is for people and planet.

A magical adventure inspired by the concept of the Circular Economy with background information on the school of thought of the Circular Economy. Plus: the title song *FOREVERCIRCLELAND* as sheet music.

MORE NOVELS TO COME...

Also by George Hohbach:

The book demonstrates how three intellectual giants of the 20th century, Kabbalist Rav Yehuda Ashlag, mathematical physicist Albert Einstein and pioneering value investor Benjamin Graham, revealed the central role of symmetry in our world and how this deep, symmetric wisdom underpins the positive, eco-intelligent and climate-smart Circular Economy.

**Follow George Hohbach on
Instagram: @georgehohbach
Twitter: @GHohbach**

POP SONGS

& YouTube Videos

Music & lyrics by George Hohbach

SYSTEM CHANGE COMPASS
GLOBAL GREEN DEAL
AGENT C2C
NATURE LIVES
C2C GALAXY
C2C CIRCULAR ECONOMY
FOREVERCIRCLELAND
WE LOVE CIRCULAR
NATURE'S SEXY CIRCLE

MORE POP SONGS TO COME...